Housing Displacement

This book examines reasons, processes, and consequences of housing displacement in different geographical contexts. It explores displacement as a prime act of housing injustice – a central issue in urban injustices.

With international case studies from the US, the UK, Australia, Canada, India, Spain, Denmark, Sweden, and Hungary, this book explores how housing displacement processes are more diverse and mutate into more new forms than have been acknowledged in the literature. It emphasizes a need to look beyond the existing rich gentrification literature to give primacy to researching processes of displacement to understand the socio-spatial change in the city. Although it is empirically and methodologically demanding for several reasons, studying displacement highlights gentrification's unjust nature as well as the unjust housing policies in cities and neighborhoods that are simply not undergoing gentrification. The book also demonstrates how expulsion, though under-researched, has become a vital component of contemporary advanced capitalism, and how a focus on gentrification has hindered a potential focus on its flipside of 'displacement', as well as the study of the occurrence of poor cleansing from a long-term historical perspective.

This book offers interdisciplinary perspectives on housing displacement to academics and researchers in the fields of urban studies, housing, citizenship, and migration studies, interested in housing policies and governance practices at the urban scale.

Guy Baeten is Professor of Urban Studies at Malmö University, Sweden, and is the Director of the newly established Institute for Urban Research. He is the principal investigator of the FORMAS Strong Research Environment CRUSH (Critical Urban Sustainability Hub). His research interests are in urban development and urban sustainability.

Carina Listerborn is a Professor in urban planning at Malmö University, Sweden. She is part of the strong research environment CRUSH (Critical Urban Sustainability Hub), and Vice-Chair of the Institute for Urban Research. Research interests are feminist urban theory, public spaces, neo-liberal planning, and housing inequalities.

Maria Persdotter is a post-doctoral researcher in welfare law at Linköping University, Sweden. Her research interests span the fields of legal geography, critical race-, and migration studies. She holds a double PhD degree in Urban Studies from Malmö University and Roskilde University.

Emil Pull is a PhD candidate in urban planning at Malmö University, Sweden. He is part of the strong research environment CRUSH (Critical Urban Sustainability Hub). Research interests are critical urban geography, critical phenomenology, and housing inequalities.

Routledge Critical Studies in Urbanism and the City

This series offers a forum for cutting-edge and original research that explores different aspects of the city. Titles within this series critically engage with, question and challenge contemporary theory and concepts to extend current debates and pave the way for new critical perspectives on the city. This series explores a range of social, political, economic, cultural and spatial concepts, offering innovative and vibrant contributions, international perspectives and interdisciplinary engagements with the city from across the social sciences and humanities.

Urban Neighbourhood Formations
Boundaries, Narrations, and Intimacies
Edited by Hilal Alkan and Nazan Maksudyan

Urban Ethics
Conflicts Over the Good and Proper Life in Cities
Edited by Moritz Ege and Johannes Moser

Urban Art and the City
Creating, Destroying, and Reclaiming the Sublime
Edited by Argyro Loukaki

Housing Displacement
Conceptual and Methodological Issues
Edited by Guy Baeten, Carina Listerborn, Maria Persdotter and Emil Pull

Mega-Events, City and Power
Nelma Gusmão de Oliveira

For more information about this series, please visit www.routledge.com/Routledge-Critical-Studies-in-Urbanism-and-the-City/book-series/RSCUC

Housing Displacement

Conceptual and Methodological Issues

Edited by
Guy Baeten, Carina Listerborn,
Maria Persdotter and
Emil Pull

Routledge
Taylor & Francis Group
LONDON AND NEW YORK

First published 2021
by Routledge
2 Park Square, Milton Park, Abingdon, Oxon OX14 4RN

and by Routledge
605 Third Avenue, New York, NY 10017

First issued in paperback 2022

Routledge is an imprint of the Taylor & Francis Group, an informa business

British Library Cataloguing-in-Publication Data
A catalogue record for this book is available from the British Library

Library of Congress Cataloging-in-Publication Data
A catalog record has been requested for this book

ISBN 13: 978-0-367-56453-7 (pbk)
ISBN 13: 978-1-138-38555-9 (hbk)
ISBN 13: 978-0-429-42704-6 (ebk)

DOI: 10.4324/9780429427046

Typeset in Times New Roman
by codeMantra

Contents

Figures

Tables

Contributors

Isabelle Anguelovski is an ICREA research professor at the Universitat Autònoma de Barcelona and co-director of the Barcelona Lab for Urban Environmental Justice and Sustainability (BCNUEJ). Trained in Urban Studies and Planning (PhD, MIT, 2011), she examines the extent to which urban plans and policy decisions contribute to more just, resilient, healthy, and sustainable cities and how community groups in distressed neighbourhoods contest the existence, creation, or exacerbation of environmental inequities as a result of urban (re)development processes and policies.

Guy Baeten, Professor of Urban Studies at Malmö University, Sweden, is the director of the newly established Institute for Urban Research. He has previously worked at the universities of Lund, Oxford, Leuven, and Strathclyde. Guy Baeten is interested in urban development projects and urban sustainability. He is the principal investigator of the FORMAS Strong Research Environment CRUSH – Critical Urban Sustainability Hub. He is involved in two research projects on smart cities with case studies in Toronto, Copenhagen, Stockholm, and Antwerp that are running between 2018 and 2020. He is also part of a project that investigates histories of the sustainability concept in Malmö urban planning.

Lisa K. Bates is an Associate Professor at the Toulan School of Urban Studies and Planning at Portland State University and a nationally recognized scholar in housing and community development on issues of gentrification and displacement. Her collaborations in Portland include working on the implementation of anti-displacement measures in the City's Comprehensive Land Use Plan and community development strategies in North/Northeast Portland. Dr. Bates has participated in this work through engaging in research partnerships and serving on technical advisory and oversight committees with the City of Portland; she maintains deep engagements with community-based organizations working towards racial justice and housing rights.

Ipsita Chatterjee is an Associate Professor at the Department of Geography, University of North Texas. She completed her PhD at Clark University,

Massachusetts. Chatterjee focuses on globalization, urban change, class, identity, Marxism, and feminism. Chatterjee's first book, *Displacement, Revolution, and the New Urban Condition* (Sage), deals with globalization and displacement of the urban poor. Her most recent book, *Religion, Landscape and the Dialectics of Globalization* (Oxford), deals with the spectacular theme park-like Disneyfication of migrant temple landscapes in Dallas and Atlanta and their counterparts in India to reveal how landscapes are narratives that ground globalization, class, racism, and sexism.

James Connolly is co-director of the Barcelona Lab for Urban Environmental Justice and Sustainability. He is also a Juan de la Cierva-Incorporacion Fellow with the Spanish Ministry in charge of research. His research – focused on urban social-ecological conflicts, urban environmental stewardship, urban social justice, urban greening, and critical urban sustainability and resilience politics – is published widely in peer-reviewed journals and books.

Louise Fabian is an Associate Professor at the Institute of Culture and Society, University of Aarhus Denmark. Her research covers the intersections between critical geography, intellectual history, cultural studies, and philosophy. She has published, researched, and taught in the areas of history of ideas, philosophy, urbanism, cultural resistance, theories of place and space, gender studies, activism, and social movements. She has been a visiting research fellow at the Centre for Cultural Studies at Goldsmiths, University of London, and a visiting research scholar at the Department of Human Geography, University of Lund, Sweden.

Melissa García-Lamarca is a postdoctoral researcher at the Barcelona Lab for Urban Environmental Justice and Sustainability (BCNUEJ), based at the Universitat Autònoma de Barcelona, and a Juan de la Cierva-Formación Fellow. She researches the financialization of housing, the lived experience of housing injustices, and how collective urban struggles can disrupt the inegalitarian status quo and open up new alternatives. At the Lab, where she is the Coordinator of Urban Housing and Justice studies, Melissa uncovers the financial dynamics of urban greening and (green) value grabbing in relation to processes of green gentrification. Melissa forms part of the core editorial collective of the *Radical Housing Journal*.

Anders Lund Hansen is a reader (docent) in human geography at Lund University and a guest professor at the Department of Philosophy and History of Ideas Aarhus University. Since 2015, he has been the academic director of the master's programme in human geography. His main research interests are political economy/ecology, urban political geography, social geography, gentrification, housing, planning, and visual methodology. He publishes widely and is a frequent speaker at international conferences. Between 2011 and 2019, Lund Hansen was an elected committee member of the International Critical Geography Group, and since 2007, he was appointed steering group member of the Christiania Researcher in Residence project.

Matt Hern is a White settler living on səlil'wətaʔɬ (Tsleil-Waututh), Skwxwú7mesh (Squamish), and xʷməθkʷəy̓əm (Musqueam) territory, with his partner and daughters. He co-directs Solid State Community Industries and continues to lecture globally. His most recent books are *What a City is For: Remaking the Politics of Displacement* (MIT Press, 2016), *Global Warming and the Sweetness of Life* (MIT, 2018, with Am Johal and Joe Sacco), and *On This Patch of Grass: City Parks on Occupied Land* (Fernwood, 2019, with Daisy Couture, Sadie Couture, Selena Couture, Erick Villagomez, Glen Coulthard, and Denise Ferreira da Silva).

Phil Hubbard is a Professor of Urban Studies at King's College London. He has broad interests in the social and cultural geographies of cities and towns, particularly the way that the socio-legal regulation of cities via municipal and environmental law constructs political and social divides between deserving and undeserving populations. His current work includes studies of coastal change and gentrification, the housing crisis in London, and the rise of vertical living. His books include *The City* (second edition, Routledge, 2018), *The Battle for the High Street* (Palgrave, 2017), and *Key Thinkers on Space and Place* (second edition, Sage, 2010). He is also currently the editor of the journal *Urban Studies*.

Loretta Lees is a Professor of Urban Geography who is internationally known for her research on gentrification/urban regeneration, global urbanism, urban policy, urban public space, architecture, and urban social theory. She has been identified as the seventeenth most referenced author in urban geography worldwide and the only woman in the top twenty (Urban Studies, 2017).

LEES, L. with Phillips, M. (eds.) (2018) Handbook of Gentrification Studies, Edward Elgar.
LEES, L., Shin, H. and Lopez-Morales, E. (2016) Planetary Gentrification, Polity Press: Cambridge.
LEES, L., Shin, H. and Lopez-Morales, E. (eds.) (2015) Global Gentrifications: uneven development and displacement, Policy Press: Bristol.
Bridge, G., Butler, T., and LEES, L. (eds.) (2011) Mixed Communities: gentrification by stealth? Policy Press: Bristol.
LEES, L., Slater, T. and Wyly, E. (2008) Gentrification, Routledge: New York.

Carina Listerborn is a Professor in Urban Planning at the Institute for Urban Research at Malmö University. She is one of the founders of the strong research environment CRUSH – Critical Urban Sustainability Hub (2014–2020), which is a national research network that puts the housing question in the centre of sustainability issues. She is also the Vice-Chair of the Institute for Urban Research (2018–). Her main research interests are feminist urban theory, public spaces, urban conflicts, neo-liberal planning, 'housing from below', and housing inequalities. Her most recent research focuses on intersectional perspectives on smart housing developments. She publishes widely and is a frequent speaker in both academic and non-academic conferences and seminars.

Anne-Sophie Iotti is a PhD student at RMIT University, Melbourne, Australia, and at Katholieke Universiteit of Leuven, Leuven, Belgium. In her PhD thesis, she investigates housing displacement in Australian cities. Prior to this, she worked as a senior research assistant at Griffith University, Brisbane, Australia, on several projects related to climate change. She is currently working at the European Research Council in Brussels, Belgium.

Jonathan McCombs is currently a PhD candidate at the Department of Geography. He also holds an MA in Sociology and Social Anthropology from Central European University. He was a Fulbright student researcher in Budapest for the academic year 2018–2019. His research examines gentrification processes in Budapest's Eighth District, where he reconsiders state-led urban renewal projects through the lens of the racial state.

Erin McElroy is a postdoctoral researcher at New York University's interdisciplinary AI Now Institute, researching artificial intelligence behind property technology. Erin is also co-founder of the Anti-Eviction Mapping Project, a data visualization, digital cartography, and multi-media collective documenting dispossession and resistance struggles upon gentrifying landscapes. Erin earned their doctoral degree in Feminist Studies from the University of California, Santa Cruz, with a focus on the politics of space, race, and technology in Romania and Silicon Valley, and continues to organize with housing justice groups in Romania. More recently, Erin co-launched the *Radical Housing Journal* to foster housing justice solidarity transnationally.

Maria Persdotter is a post-doctoral researcher in welfare law at Linköping University and a part-time research affiliate at the Institute for Studies in Malmö's History at Malmö University. Originally trained as a human geographer, she holds a double PhD degree in Urban Studies from Malmö University and Roskilde University. Her PhD Dissertation, entitled *Free to Move Along*, examines the urbanization of bordering practices against impoverished Roma EU citizens through a case study of the displacement of a squatter community in Malmö, Sweden. Her research interests span the fields of critical race, migration, and urban studies. Currently, she is involved in a project on the changing interpretations of the right to social assistance for irregular migrants in Sweden.

Emil Pull is a PhD candidate in urban studies at Malmö University, Sweden, and at the University of Roskilde, Denmark. Emil's thesis work focuses on urban transformation and displacement following large scale renovations in Uppsala, Sweden.

Chiara Valli has a PhD in Social and Economic Geography from Uppsala University. After a postdoc at the University of Gothenburg, she is currently employed as a postdoc researcher at the Institute for Urban Research, Malmö University within the CRUSH (Critical Urban Sustainability Hub) research platform. As an urban geographer interested in the social and political dimensions of urban restructuring and their impacts on social sustainability and urban (in)equality, her research includes issues of gentrification, displacement, segregation, labor and precarity, urban activism, participatory and visual research methodologies. She has conducted fieldwork in Italy, US, and Sweden.

Acknowledgements

The idea to write this book has gradually grown out of our research activities in the Swedish Strong Research Environment 'Critical Urban Sustainability Hub' or CRUSH. At the Nordic Geographers Meeting (NGM) in Stockholm in 2017, we organized a paper session under the title 'Beyond displacement: conceptual and methodological issues'; most of the chapters in this book are based on these presentations. We want to thank the conference organizers Gunnel Forsberg and Jennie Gustafsson at Stockholm University, who accommodated the large CRUSH crew at this event. We are grateful to all contributors to the conference and to this book for helping us advance the thinking around housing displacement.

We especially want to thank our reviewers, who unconditionally engaged with the chapters of this book. Your comments and input have been invaluable. Thank you, Richard Ek, Johan Pries, Helena Holgersson, Erik Jönsson, Carl Cassegård, Claudia Fonseca Alfaro, Özlem Çelik, Catharina Thörn, Jennie Gustafsson, Linda Lapina, and Chiara Vitrano. CRUSH was funded by the Swedish research council FORMAS (no. 250-2103-1794).

Introduction

Housing displacement: conceptual and methodological issues

Emil Pull, Guy Baeten, Carina Listerborn, and Maria Persdotter

It is in the act of displacement that housing injustice finds its prime expression. Therefore, displacement needs to take a much more central place in our understanding of urban injustices. With this book, we want to reveal how housing displacement processes mutate into new forms and are more diverse than have been acknowledged thus far in the literature. We need to think beyond the existent gentrification literature to understand the reasons and consequences of housing displacement, especially considering geographical differences.

We propose giving primacy to researching displacement processes when studying socio-spatial change in the city – even though it is empirically and methodologically more demanding – for several reasons. First, through studying displacement we can, as said, put more focus on gentrification's unjust nature. Second, we can highlight unjust housing policies in cities and neighbourhoods that are simply not undergoing gentrification. Third, it is vital to demonstrate how expulsion in all its varieties, though under-researched, has become a vital component of contemporary advanced capitalism (Sassen, 2014). Fourth, the dominance of the concept of gentrification has not only hindered a potential focus on its flipside – displacement – but it has also hindered the study of the occurrence of poor cleansing from a long-term historical perspective. While gentrification is still taking place and will take place in many cities across the globe, it should perhaps no longer be considered the *defining factor* of unjust urban social change – displacement precedes gentrification. We suggest that the systematic ways in which governments and landlords seek to remove unwanted segments of the population from (inner) cities or prevent them from moving in, with or without subsequent gentrification, is the defining factor of contemporary urban injustice. Thus, this book not only gives examples of how displacement manifests itself in different contexts, it also aims to further our understanding of the underlying power relations.

Displacement as a concept and a reality

Urban displacement has a long history. Removing the poor and banning outcasts from cities has been a systematic policy practice since at least medieval times, when ordinances regulated who could be in the city and who could not (Coy, 2008). In an attempt to minimize the expenditures for poor relief and imprisonment, and instil some kind of urban order, hundreds of thousands of beggars, petty thieves, 'gypsies', and other marginalized groups were systematically transported from imperial England to, first, the Americas and later to Australia (an act of displacement that presumed and further compounded the displacement of indigenous peoples from those same lands). If anything, poor cleansing, followed by their replacement with a well-off population or not, has been a pivotal element of urban social history for centuries (Beier, 1985; Himmelfarb, 1985; Cresswell, 2001; Beier and Ocobock, 2008). In *The Housing Question*, Engels (1872: 18) writes about Parisian workers who 'are forced out of the centre of the towns towards the outskirts' and how their 'dwellings, and small dwellings in general, become rare and expensive and often altogether unobtainable'.

Within the realm of urban studies, Eunice Grier and Georg Grier (1978) are usually considered the first to give displacement a feasible definition. Following congressional concerns over displacement in the United States, Grier and Grier were charged by the Department of Housing and Urban Development to investigate the scope and ramifications of what was seen as the 'negative and highly charged by-product of what is otherwise a most helpful (and largely unexpected) development for America's central cities' (Grier and Grier, 1980: 252). Taking a broad approach, they surmised that displacement is not solely an urban issue but prominent in rural communities as well and that '[t]he responsible agent is not always a speculator or private "rehabber". It can be a public agency, a quasi-public or private institution, or even the forces of nature' (ibid.: 253). Further, '[t]he households displaced are not all poor or members of minorities. They can be renters and owners, they are of all races, and they span a considerable range of ages, incomes and household types' (ibid.). In this broad take, the authors list twenty-five factors leading to displacement, including accidental fire, foreclosure, highway construction, urban renewal, natural disaster, and renovation of public housing. The list implies a diverse set of actors: natural phenomena, private investors, government planning decisions, federal urban renewal schemes, etcetera. In a capture-all definition, Grier and Grier (1980: 256) conclude the following:

> Displacement occurs when any household is forced to move from its residence by conditions which affect the dwelling or its immediate surroundings, and which:
>
> 1 are beyond the household's reasonable ability to control or prevent;
> 2 occur despite the household's having met all previously imposed conditions of occupancy; and
> 3 make continued occupancy by that household impossible, hazardous, or unaffordable.

Importantly, Grier and Grier make no clear distinction between voluntary and involuntary displacement. Adhering to the multitude of subtle actions, tactics, and inactions leading up to displacement, they argue,

> For most residents to move under such conditions is about as 'voluntary' as is swerving one's car to avoid an accident. By the time the landlord issues notices of eviction, or the code inspector posts the structure as uninhabitable, few occupants may be left. Therefore we cannot define displacement simply in terms of legal or administrative actions – or even draw a clear-cut line between 'voluntary' and 'involuntary' movement.
> (Ibid.: 253)

A few years later, Hartman, Keating, and LeGates published the anti-displacement handbook *Displacement, how to fight it*, where they define displacement as 'what happens when forces outside the household make living there impossible, hazardous, or unaffordable' (1982: 3). This simple definition has been a go-to definition for many displacement studies since. Another landmark study that built on the efforts of Grier and Grier (1980) was Peter Marcuse's (1985) investigations of displacement occurrences in New York. His study was very much entangled with the contemporaneous gentrification debate and took displacement as the outcome of changing land values, resultant of cyclical relations between abandonment, investment, and gentrification. Marcuse's view of gentrification-induced displacement as the outcome of cycles of inner-city investment and divestment was very timely and held a high degree of explanatory power in the United States at the time. But as Elliott-Cooper et al. (2019) note, it might not always speak to the twenty-first-century experiences of urban displacements around the world. The conceptual legacy of Marcuse's text today might lie outside of the economic and political mechanisms that drove those processes in the 1980s United States. For Marcuse, the intertwinement of abandonment and gentrification and the multifaceted process of neighbourhood change called for a revision of the displacement concept. Only last-residence displacement might be methodologically convenient, but it does not account for the indirect displacement of households as they, over time, get barred from certain areas and neighbourhoods due to gentrification and increasing rents. Confronting Grier and Grier, Marcuse consequently conceptualised four forms of displacement: *direct last resident displacement* (over either economic reasons, or coercion by the landlord), *direct chain displacement* (looking at tenants that have moved in an earlier stage of decline or investment, or previous rent increases), and *exclusionary displacement.* Regarding the latter, Marcuse says (adding to Grier and Grier's definition of displacement),

> Exclusionary displacement from gentrification occurs when any household is not permitted to move into a dwelling, by a change in conditions, which affect that dwelling or its immediate surroundings, which:
>
> 1 is beyond the household's reasonable ability to control or prevent;
> 2 occur despite the household's being able to meet all previously imposed conditions of occupancy;

3 differs significantly and in a spatially concentrated fashion from changes in the housing market as a whole; and
4 makes occupancy by that household impossible, hazardous, or unaffordable.

(1985: 206p)

The fourth form of displacement is *displacement pressure*, which Marcuse conceptualises as follows:

> [D]isplacement affects more than those actually displaced at any given moment. When a family sees the neighborhood around it changing dramatically, when their friends are leaving the neighborhood, when the stores they patronize are liquidating and new stores for other clientele are taking their places, and when changes in public facilities, in transportation patterns, and in support services all clearly are making the area less and less livable, then the pressure of displacement already is severe. Its actuality is only a matter of time. Families living under these circumstances may move as soon as they can, rather than wait for the inevitable; nonetheless they are displaced. This is true both for displacement from gentrification and for abandonment.
>
> (Ibid.: 207)

While these sophisticated definitions by Marcuse came to guide the displacement literature, displacement as a studied phenomenon was still often reduced to last-resident outmigration in the empirical studies that followed. Another consequence of Marcuse's and other early displacement studies from the 1980s was that displacement, at least within the field of urban studies, got entangled with and seen as the backside of gentrification. Because of this entanglement, it is difficult to discuss displacement without working through how it has been dealt with in the evolution of gentrification as practice and theory.

Displacement and gentrification

Coined by the Marxist urban scholar Ruth Glass in the 1960s, the tongue-in-cheek term *gentrification* was used to describe how working-class neighbourhoods in London were invaded by an upper- and lower-middle-class gentry. She concluded that '[o]nce this process of "gentrification" starts in a district it goes on rapidly, until all or most of the original working-class occupiers are displaced and the whole social character of the district is changed' (Glass cited in Smith, 2002: 438). However, just as the history of urban displacement is old, gentrification too has claims to an older history. Returning again to Engels, his characterization of Paris bears a striking resemblance to the processes Glass describes:

> By 'Haussmann' I mean the practice, which has now become general, of making breaches in the working-class quarters of our big cities,

> particularly in those which are centrally situated, irrespective of whether this practice is occasioned by considerations of public health and beautification or by the demand for big centrally located business premises or by traffic requirements, such as the laying down of railways, streets, etc. No matter how different the reasons may be, the result is everywhere the same: the most scandalous alleys and lanes disappear to the accompaniment of lavish self-glorification by the bourgeoisie on account of this tremendous success, but – they appear again at once somewhere else, and often in the immediate neighbourhood.
>
> (Engels, 1872: 70)

Engels' 'Haussmann' (from the Bonaparte urban planner Georges-Eugène Haussmann, who oversaw the rebuilding of Paris with modern avenues and open spaces – a process that is estimated to have displaced 350,000 people) is indeed very reminiscent of the early conceptualizations of gentrification. The understanding of urbanization as a class formation of space is as central to Engels' 'Haussmann' as it was to Ruth Glass' (1964) gentrification in London, and later to the prominent gentrification scholars of the 1980s (LeGates and Hartman, 1982; Smith, 1982; Marcuse, 1985). Already from its inception, displacement was at the fore of gentrification analysis and treated as the inevitable flipside of gentrification.

The concept of gentrification caught on quickly in an Anglo-American context, and it rapidly became an important analytical tool wielded by both radical academics and activists as a critique of a host of different urban projects. The apparent similarities in the process and outcome of various urban renewal processes through time and space have since kept gentrification as a continuous and prominent trope in critical urban research. However, the early investigations into the lives of the displaced and the critical reading of gentrification in the literature of the 1970s and 1980s (Hartman, 1979; Hartman et al., 1982) gave way to several crises for gentrification studies in the late 1980s and 1990s. The first crisis was related to a short-lived expectation that gentrification as a phenomenon would diminish in scope in the backwater of the economic recession and be relegated to the margins (Bourne, 1993). The second was a conceptual debate over the existence and explanatory power of rent gap theory (Bourassa, 1993; Clark, 1995). The third was a polarized debate between two sides: Marxist political economy, which saw gentrification as supply-driven and operating squarely in the realm of capitalist production and exchange (with uneven property relations at the centre), and cultural explanations, which saw gentrification primarily as consumer-driven by a new bourgeoning middle class and their consumption preferences. The decidedly Anglo-American legacy of gentrification research led to gentrification being seen as a back-to-the-city movement, and it is not surprising that gentrification research gained traction when and where it did (i.e., during the late 1970s and 1980s in the great Anglo-American metropolises). Planners and capital saw gentrification (or re-investment/revitalization) as the solution to a period of inner-city decline, particularly in the United States. The middle class that had left the American cities for the suburbs was to be lured back through investments into the physical environment and through increased property values. For some (e.g., Ley, 1980), the back-to-the-city movement was driven by changes in consumer preferences following the advent of the post-industrial city. For others (Smith, 1979; Marcuse, 1985; Clark, 1988), it was a decidedly supply-driven affair, and the back-to-the-city movement was first and foremost the movement of capital by investors that saw potential profits. Clark (1988: 86) also argues that 'in their empirical form at the level of appearance, the results may be interpreted from either perspective'. This lively debate influenced a great deal of gentrification research over the following decades. However, by the end of the 1990s, the heated debate of the 1980s and early 1990s had given way to a declining interest in the phenomenon of gentrification by researchers (Lees, 2000).

At the same time, the prominence given to cultural explanations and consumer preferences in driving gentrification meant that displacement became less of a concern in gentrification studies[1] (Hamnett, 1991, 2003; Vigdor, 2002, 2010; Freeman, 2005; Butler, 2007; McKinnish et al., 2010; Kohn, 2013; Freeman et al., 2016; Meltzer and Ghorbani, 2017). The 2000s and 2010s marked a shift towards what Wacquant (2008) has labelled the 'gentrification of gentrification' and Slater (2006) 'the eviction of critical perspectives in gentrification research'. The middle class became the protruding subject in many gentrification studies, at the cost of working-class perspectives. This erasure of working-class perspectives led Helbrecht (2018) to describe gentrification research as a one-eyed cyclops operating with an intellectual bias, since it has become mostly concerned with the upgrading of a neighbourhood and not with the displaced populations.

Nevertheless, while displacement became less of an issue in some of the literature of gentrification, critical scholars have brought gentrification, both empirically and conceptually, to new terrains. This is perhaps most clearly epitomized in the globalized/planetary gentrification and rent gap theses of authors such as Smith (2002), Slater (2006), Wyly (2015), Ley and Yang (2017), and Lees et al. (2018). It would be a gross oversimplification and injustice to read their contributions to mean that gentrification is taking the same forms across the world. On the contrary, their effort highlights the variegated shapes and forms gentrification can take. Undeniably, efforts at restructuring urban spaces for the use of more affluent groups is not isolated to Britain or the United States. Scholars have described a wide set of processes such as slum demolition in China, India, and Pakistan and mass displacement in Bangladesh, India, and Malaysia as gentrification (Lees, 2012). Some scholars consider this conceptual expansion fruitful for consolidating solidarity across diverse and disparate accounts of displacement, building collective identities and belonging (Butler, 2007), and rendering itself useful to global epistemologies of the urban, such as Neil Brenner and Christian Schmidt's planetary urbanization project (Brenner and Schmid,

2015). Nevertheless, it has received criticism for lacking in contextual sensitivity and for constituting a colonial imposition of North-centrist theorization in settings better understood using a different lexicon. For instance, Ghertner (2014) and Maloutas (2012, 2018) propose that the 'catch-all' use of gentrification threatens to obscure and cloud the diverse empirical dynamics behind various displacement processes, embedded as they are in different systems of property and planning. Maloutas (2012) argues that the contextual baggage of gentrification leads to losses in analytical rigour, and Ghertner (2015) means that capturing the diversity of displacement logics under the same umbrella will in the worst case produce improper conclusions. This debate is still ongoing and constitutes one of the more lively and fruitful debates in gentrification studies.

Displacement sans gentrification

There is reason to question this conventional canon of urban displacement, and to perhaps not change but at least supplement its genealogy. Another viable reading of displacement could start with Fried's (1964) study 'Grieving for a lost home: Psychological cost of relocation'. This phenomenologically astute study into the feelings of displaced persons following a slum clearing in Boston has been a staple reference for a strand of displacement research that has not been as tied to the investment-divestment paradigm of gentrification research. In recent years, this literature has been 'rediscovered', and research into the experiences of displaced populations is increasing within urban studies (Davidson, 2009; Atkinson, 2015; Valli, 2015; Pull and Richard, 2019). However, these investigations, with their focus on *dwelling and being*, have a history of their own outside of the mainstream gentrification paradigm. By way of example, Thursz draws directly on Fried's concern for feeling-at-home and the traumatic experience of relocation as a consequence of neighbourhood destruction, when he already in 1966 writes of a slum clearance in Washington DC and asserts,

> No matter how dirty, inadequate, and unsanitary the old Southwest was, it was also home for families that had been there for a long time [...] For many, the loss was deep and continues to be felt. They have not "readjusted" and one-fourth of the relocatees have not made a single friend since leaving the old Southwest.
>
> (Thursz, 1966: 101–102)

Phenomenologically inspired places-as-lived-spaces studies drawing on insights from Fried (1964), Relph (1976), Tuan (1971, 1979), Heidegger and Hofstadter (1971), and Lefebvre (1991), and other place theorists have deepened our understanding of home, neighbourhood, and the nature of displacement. The process of displacement from a phenomenological standpoint is not primarily a question of relocation that can be boiled down to (more or less forced) mobility on the Euclidian plane. Relegating displacement to the realm of relocation in abstract space, as movement from point A to point B, misses important place–space tensions. In Davidson's (2009: 223) words, 'it reduces a socio-spatial phenomena to a purely spatial event [...] Put simply, displacement understood purely as spatial dislocation tells us very little about why it matters'. For Davidson, it is a point of view that frames a complex social phenomenon as a simple question of the 'to be or not to be' of physical relocation. However, phenomenology has taught us that displacement processes are perhaps primarily about the loss of lived spaces – whether or not gentrification is the precursory process or physical relocation is what follows. Many cases of displacement can perhaps more aptly be described as 'un-homing' (Atkinson, 2015) or 'domicide' (Porteous and Smith, 2001); the intentional destruction of home; or a root shock (Fullilove, 2001, 2005), which is a deeply disruptive process born out of the destruction of social networks and community, place attachment, and history.

There is also a growing and immensely rich body of ethnographic work on displacement. For instance, Desmond's *Evicted* (2016) only very rarely and hesitantly talks about gentrification, but its account of the lives of evicted and displaced persons in Milwaukee makes one of the stronger arguments for why housing and poverty always need to be discussed alongside each other. Hern's *What a city is for* (2016) explores the racialized displacement of black communities and neighbourhoods in some of the richest cities in North America and argues that in resisting displacement, one will always have to question the very way development, private property, and land functions is understood. Desmond and Hern, in their own way, repeat what Engels (1872) took to be true already in his time: Poverty, housing precarity, and displacement are not unfortunate by-products of more or less benign, or more or less malignant, types of urban interventions and urbanism. Rather, they are a functional and integral part of how we organize our capitalist cities. Displacement then is not only the backside of gentrification but the backside of the capitalist mode of city production itself – infused by naturalized notions of ownership that often operate along racist, gendered, and classist logics, entwined in the western history of property and the history of land theft of indigenous land (see Bates and Hern in Chapter 11).

Nevertheless, even if scholars see displacement – as we do – as foundational to the way cities are produced and reproduced, displacement plays out differently in different parts of the world, in different times, and in different contexts. Consequently, there can be no catch-all definition of displacement, not one set of concepts to capture it, and not one way of studying it in the field. On the contrary, due to the sometimes-elusive nature of displacement, the difficulty in deriving its scale,[2] and the varying contexts in which it occurs, methodological advances are at least as paramount as conceptual ones. We are therefore sympathetic to efforts such as González (2016) call for critical comparative work in studying displacement.

Thrust and outline of the book

To summarize the thrust of the book, we believe a significant enrichment of the concept of displacement is necessary – for at least three reasons. First, displacement research came about in the context of gentrification and has ever since been intimately linked with gentrification research and treated as the 'flipside' of gentrification. However, our case studies show that different forms of displacement are taking place *without* gentrification processes. Governments and private actors are pursuing policies and strategies to remove lower-income segments without complementary 'gentrification' plans in place. It appears as if displacement has become a goal in itself – a strategy to make the city more affluent and appealing not by attracting the rich but by removing the poor. Therefore, we feel the need to think 'displacement' without linking it to the rich gentrification literature. This is not to say that gentrification is no longer an important source of displacement but that it has been accompanied by other displacement processes unrelated to gentrification. Second, the dominant strand of displacement research as we know it came about in a US context. Displacement processes with similar outcomes unfold in very different geographical contexts, from rapidly expanding cities in the Global South to North-European cities where (the remnants of) the welfare state often provide a certain protection, albeit not watertight, against actual evictions or homelessness. For example, evictions are rare in Sweden, but displacement can take very subtle (yet effective) manifestations that make us consider whether other concepts would be more adequate to describe 'displacement' in that context. Third, displacement research as we know it came about in the 1970s. Since then, the housing market, the nature of urban renewal projects, the actors on the housing market, income structures, etcetera have all undergone significant changes, leading to new processes of displacement that are not easily captured through 'traditional' understandings of displacement.

The experience of alienation in modern cities is that of precarity, insecurity, and disempowerment and increased inequalities; it is produced by tangible political and material change and upheld by tangible actors like landlords, banks, municipalities, developers, and other social actors. Some of these experiences are the result of gentrification processes, some are the result of displacement-as-outmigration processes. Yet others are not. The victims of the displacement process that precedes gentrification are often invisible, untraceable, and unmeasurable: displacees are notoriously difficult to research (Atkinson, 2000; Newman and Wyly, 2006). By giving primacy to researching displacement (whether followed by gentrification or not), we argue we can better highlight the social injustice implied in the process. Some have argued that any gentrification study that does not take displacement seriously can be regarded as an incomplete enquiry; nevertheless, despite the long history of gentrification research and enquiries into displacement from various disciplines, our understanding of displacement remains limited (Slater, 2006; Shin, 2018). As Marcuse (2010: 1987) states,

> If the pain of displacement is not a central component of what we are dealing with in studying gentrification – indeed, is not what brings us to the subject in the first place – we are not just missing one factor in a multi-factorial equation; we are missing the central point that needs to be addressed.

The eleven chapters in this book cover eleven different urban contexts of displacement processes on four different continents. The chapters deploy different theoretical, empirical, and methodological perspectives, but together they show the complexity of researching, resisting, and understanding housing displacement.

Phil Hubbard and Loretta Lees discuss many of the methodological difficulties of capturing and measuring displacement in their investigations into the renewal of London's council estates. They emphasize that the displacement associated with renewal is rarely a one-off event but more often 'a form of slow, percolating violence' (Chapter 1) akin to Marcuse's displacement pressures. The lack of robust data and methodological hurdles makes it increasingly difficult to reveal the actual flows of displaced people from the estates and to differentiate between voluntary and involuntary movement. Instead, Hubbard and Lees had to rely on proxy measures to assess the degree of displacement. Their chapter emphasizes the need to create or make available longitudinal sources of data at the individual level; otherwise, 'the extent of residential gentrification-induced displacement will remain largely unrecorded and invisible' (Chapter 1) and therefore difficult to politicize and contest.

In Ipsita Chatterjee's 'conceptual musings' on her research in Ahmedabad, she raises the issue of the experienced inevitability of people threatened to be displaced. The informal settlement on the banks of the Sabarmati is one of the bases for the working poor. In the name of modernization and urban renewal, the riverbank residents were to be resettled further away from the centre, which hampers both their ability to make a living and disrupt social orders based on religious groupings. Going back to her research in Ahmedabad, Chatterjee reflects on the need for self-conceptualizations based on listening to the displaced. She learnt that people do not care why they are being moved, which raises the need for conceptualization. Through the method of listening to the displaced, the power relations of class, race, religion, and spatial contradictions come in the forefront, and this is where Chatterjee believes the entry point for understanding displacement should be. Chapter 2 illustrates the complexities and realities that underlie conceptualizing and empirically grasping displacement processes. It is a call to make our research relevant to the displaced and to the processes we research.

Chapter 3 by Emil Pull revolves around a case study of an ongoing large-scale renovation of two Swedish neighbourhoods in Uppsala – a process that is driving up rents, sometimes by as much as sixty to a hundred per cent. Because there has been little in-migration of wealthier households since the renovations started, Pull asks if it really makes sense to conceptualize this urban renewal effort as a case of gentrification. Discussing the different dimensions of and methods to capture gentrification, Pull makes the conceptual argument that what is happening in Uppsala is best characterized as a form of 'post-gentrification'. Namely, the rent-gap is being closed not by replacing the existing population with a wealthier (and more White) one but by off-loading costs on the existing tenants – and on the municipality, which is forced to increase its housing benefits.

In Chapter 4, Chiara Valli complements the existing empirical and theoretical approaches in displacement research by focusing on emotional displacement and the experience of loss of place. Valli aims to understand the reactions to, reasons for resistance of, and feelings towards urban renewal projects that trigger displacement pressure. Based on interviews in Bushwick in New York City, Valli finds that the interviewees' experienced misrecognition and fear of losing their own history and place turn into anger and resentment towards the gentrifiers – a White, arty middle class. The interviews illustrate a communicative divide, filled with symbolic power, between the newcomers and the long-time residents. The divide grew out of different experiences of and roots in the place. Valli's article shows the importance of understanding displacement pressure on a social and cultural level and how this pressure slowly changes social power relations alongside the expected rent and land value increases.

The District Mayor of the inner-city area in Budapest, Mate Kocsis, initiated an urban development process at the residence site of the largest Roma community in the city. The aim was not only to lift the place physically but also to include the residents themselves through, for example, vocational training, social programs, and crime prevention. In Chapter 5, Jonathan McCombs read this as a biopolitical process that enrols 'the urban fabric into a state-led project of population management'. Focusing on the post-socialist housing market, which is different from the markets of Europe and the United States, McCombs develops a theoretical framework to understand how hybrid capitalist-socialist property relations are weakened by racialized state-led interventions. Here, the Roma people were the target of both displacement and cultural adaptation. The biopolitical project by the government to intervene in the life and culture of the existing Roma population was at the forefront of the urban project.

Erin McElroy, in Chapter 6, shares empirical examples from San Francisco that illustrate how shell companies, protected by state laws, force evictions of rent-controlled tenants to transform rental housing into condominiums. These companies speculate on the housing market under strategic anonymous labels, which makes it virtually impossible for tenants to know their counterpart and thereby decreases the possibilities to raise resistance to the evictions. McElroy highlights the raw consequences of an increased speculative real estate market where dispossession processes are performed by an 'invisible' financial hand. Further, these financial strategies act upon racialized and colonialized land, reinforcing existing inequalities. Based on the Anti-Eviction Mapping Project, McElroy adds hope of resistance by pointing at the importance of activism and research to reveal the strategies and consequences of 'undercover' real estate actors.

What motivates the government of the mid-sized town Landskrona in the south of Sweden to speak out against people living on social benefits? And what measures were taken to steer people on benefits away from the city? Baeten and Listerborn give an example in Chapter 7 of a slow but persistent state-led process to change the social fabric and the housing market of a city. Through an urban policy with a focus on the real estate market, the City of Landskrona initiated changes to make it more difficult for poor people to stay in the city centre. In an attempt to reveal the step-by-step procedures, and the actual changes, this case provides the grounds to discuss banishment as an urban renewal strategy. Banishment captures these subtle policies that will ultimately lead to significant changes to the city's social-demographic fabric. Banishment is not just one strategy; rather, it is based on a repertoire of policies. The city's 'nudging tactics' makes it difficult to reveal the actual consequences and to find victims and possibly mobilize resistance. Perhaps this is why it is so effective as an urban renewal strategy.

Chapter 8 by Louise Fabian and Anders Lund Hansen sheds light on the Danish housing market and the current mobilization around resistance to displacement in the face of the privatization of non-profit housing and the increasingly racist discourse around segregation. The Danish government's so-called ghetto plan, which aims to turn around stigmatized neighbourhoods and 'terminate the ghettos completely', has energized housing activism around the right to the city, housing, and freedom from displacement. The chapter contextualizes resistance with respect to several successive attempts to commodify Danish public housing. In so doing, this chapter highlights a paradox: whereas ghettos have been a government technology used for containing racialized minority groups (originally mainly Jewish populations), the Danish government is using the language of the ghetto to enforce an aggressive policy of dispersal.

Brisbane, the third largest city of Australia, has since the 1990s undergone an intensified transformation with the aim of urban consolidation. In Brisbane, major renovations took place in old working-class neighbourhoods. Anne-Sophie Iotti describes in Chapter 9 the process of displacement that sets off when cities with low density, like Brisbane, face new policies to consolidate and densify. The infill redevelopment in major cities and along major transport corridors have reversed the previous trends of population decline in inner cities. The inner city of Brisbane is now 'attractive' because of the explicit strategy to attract high-income households without dependent children, namely, young professionals. Amenities like museums, art galleries, and restaurants have been improved for the people living in the renovated central locations. In contrast with the existing housing stock, the developers are building small apartments with luxury standards to increase profit. Iotti describes how displacement is being normalized and regarded as unavoidable by the urban developers and government. In interviews with displaced residents, she identifies the consequences of lack of maintenances, shortage of affordable housing, lack of protections for renters, and neighbourhood changes.

In Chapter 10, Melissa García-Lamarca, James Connolly, and Isabelle Anguelovski turn to green planning and its consequences for processes of gentrification and displacement, through mapping greening projects in Barcelona since the 1990s. What might appear as investments in sustainable planning carry the risk of making some areas more attractive than others with the result of initiating a gentrification process. However, green projects may also form a threat to displaced homeless groups and increase urban segregation. Through the case study in Barcelona, it becomes clear that wealthy areas are benefitting more from green investments than the dense and impoverished inner-city areas.

Finally, in Chapter 11, Matt Hern and Lisa K. Bates open up for a wider discussion around displacement, where the solution for 'dispossession' does not necessarily have to be 'possession' (as Judith Butler phrased it). Learning from the history of North America, 'Black labour on North American Indigenous land' sheds a critical light on the issue of (White) property. What are the possibilities to not only think but also act in a way that breaks with existing processes of exclusion and appropriation? Hern and Bates' dialogue seriously challenges the problem formulation of housing displacement and invites formulating an agenda for a more equal housing market.

Notes

1 Based on a review of 114 empirical studies of gentrification between 1973 and 2001, Atkinson (2002) notes how displacement is the most dominant theme in the literature, covering more than half (71) of the examined studies. It is also clear that displacement recedes as the topic of inquiry towards the 1990s and early 2000s. Over the years 1973–1985, 37 out of 41 studies (ninety per cent) examined gentrification-induced displacement. For the years 1986–2001, the figure is 34 out of 67 (fifty per cent).

2 Zuk et al. (2015) note how so many factors affect the derived 'scale' of the problem: data availability, the deployed statistical methods, and the timing of the study, coupled with the questions the study sets out to answer. The latter include the nature of displacement (e.g. how many are displaced, where they go, how it varies between groups, etc.), the causes (e.g. rent changes, disinvestment, etc.), or the consequences (e.g. re-segregation, crowding, rent-burden, neighbourhood destabilization, etc.).

References

Atkinson, R. (2000). Measuring gentrification and displacement in Greater London. *Urban Studies*, 37(1), pp. 149–165.

Atkinson, R. (2002). Does gentrification help or harm urban neighbourhoods? An assessment of the evidence-base in the context of the new urban agenda (CNR paper 5). *ESRC Centre for Neighbourhood Research* (CNR Paper), (June), p. 27. Available at: http://eprints.whiterose.ac.uk/46553/

Atkinson, R. (2015). Losing one's place: Narratives of neighbourhood change, market injustice and symbolic displacement. *Housing, Theory and Society*, 32(August), pp. 1–16. doi:10.1080/14036096.2015.1053980.

Beier, A. L. (1985). *Masterless men: The vagrancy problem in England 1560–1640*. London: Methuen.

Beier, A. L. and Ocobock, P. (eds.) (2008). *Cast out: Vagrancy and homelessness in global and historical perspective*. Athens: Ohio University Press.

Bourassa, S. C. (1993). The rent gap debunked. *Urban Studies*, 30(10), pp. 1731–1744.

Bourne, L. S. (1993). The myth and reality of gentrification: A commentary on emerging urban forms. *Urban Studies*, 30(1), pp. 183–189. doi:10.1080/00420989320080111.

Brenner, N. and Schmid, C. (2015). Towards a new epistemology of the urban? *City*, 19(2–3), pp. 151–182. doi:10.1080/13604813.2015.1014712.

Butler, T. (2007). For gentrification? *Environment and Planning A*, 39(1), pp. 162–181. doi:10.1068/a38472.

Clark, E. (1988). The rent gap and transformation of the built environment: Case studies in Malmö 1860–1985. *Geografiska Annaler. Series B, Human Geography*, 70(2), pp. 241–254.

Clark, E. (1995). The rent gap re-examined. *Urban Studies*, 32(9), pp. 1489–1503. doi:10.1080/00420989550012366.

Coy, J. P. (2008). *Strangers and misfits: Banishment, social control, and authority in early modern Germany*. Leiden: Brill.

Cresswell, T. (2001). *The tramp in America*. London: Reaktion Books.

Davidson, M. (2009). Displacement, space and dwelling: Placing gentrification debate. *Ethics, Place and Environment*, 12(2), pp. 219–234. doi:10.1080/13668790902863465.

Desmond, M. (2016). *Evicted*. London: Penguin.

Elliott-Cooper, A., Hubbard, P., and Lees, L. (2019). Moving beyond Marcuse: Gentrification, displacement and the violence of un-homing. *Progress in Human Geography*. doi:10.1177/0309132519830511.

Engels, F. (1872). *The housing question*. Moscow: Progress Publishers.

Freeman, L. (2005). Displacement or succession? Residential mobility in gentrifying neighborhoods. *Urban Affairs Review*, 40(4), pp. 463–491. doi:10.1177/1078087404273341.

Freeman, L., Cassola, A., and Cai, T. (2016). Displacement and gentrification in England and Wales: A quasi-experimental approach. *Urban Studies*, 53(13), pp. 2797–2814. doi:10.1177/0042098015598120.

Fried, M. (1964). Grieving for a lost home: Psychological costs of relocation. In: Wilson, J. Q. (ed.) *Urban renewal: The record and the controversy*. Cambridge: The MIT Press, pp. 359–379.

Fullilove, M. T. (2001). Root shock: The consequences of African American dispossession. *Journal of Urban Health*, 78(1), pp. 72–80. doi:10.1093/jurban/78.1.72.

Fullilove, M. T. (2005). *Root shock: How tearing up city neighborhoods hurts America, and what we can do about it*. London: Random House.

Ghertner, D. A. (2014). India's urban revolution: Geographies of displacement beyond gentrification. *Environment and Planning A*, 46(7), pp. 1554–1571. doi:10.1068/a46288.

Ghertner, D. A. (2015). Why gentrification theory fails in 'much of the world'. *City*. Taylor & Francis, 19(4), pp. 552–563. doi:10.1080/13604813.2015.1051745.

Glass, R. (1964). *London: Aspects of change*. London: MacGibbon & Kee.

González, S. (2016). Looking comparatively at displacement and resistance to gentrification in Latin American cities. *Urban Geography*, 37(8), pp. 1245–1252. doi: 10.1080/02723638.2016.1200337.

Grier, G. W. and Grier, E. S. (1978). *Urban displacement: A reconnaissance*. Grier Partnership. Available at: https://books.google.se/books?id=yBDMSAAACAAJ

Grier, G. W. and Grier, E. S. (1980). Urban displacement: A reconnaissance. In: Laska, S. B. and Spain, D. (eds.) *Back to the city: Issues in neighborhood renovation*. New York: Pergamon Press, pp. 252–268.

Hamnett, C. (1991). The blind men and the elephant: The explanation of gentrification. *Transactions of the Institute of British Geographers*, 16(2), pp. 173–189.

Hamnett, C. (2003). Gentrification and the middle-class remaking of inner London, 1961–2001. *Urban Studies*, 40(12), pp. 2401–2426. doi:10.1080/0042098032000136138.

Hartman, C. (1979). Comment on 'Neighborhood revitalization and displacement: A review of the evidence'. *Journal of the American Planning Association*, 45(4), pp. 488–491. doi:10.1080/01944367908976994.

Hartman, C. W., Keating, W. D., and LeGates, R. T. (1982). *Displacement, how to fight it*. National Housing Law Project. Available at: https://books.google.se/books?id=0wIPAQAAMAAJ

Heidegger, M. and Hofstadter, A. (1971). Building dwelling thinking. *Poetry, Language, Thought*, (1), p. 42. doi:CBNotes.Card.

Helbrecht, I. (ed.) (2018). *Gentrification and resistance: Researching displacement processes and adaption strategies*. Wiesbaden: Springer. doi:10.1007/978-3-658-20388-7_7.

Hern, M. (2016). *A what a city is for: Remaking the politics of displacement*. Cambridge, MA: The MIT Press, 2016. Available at: http://proxy.mah.se/login?url=http://search.ebscohost.com/login.aspx?direct=true&db=cat05074a&AN=malmo.b2077496&site=eds-live

Himmelfarb, G. (1985). *The idea of poverty*. New York: Random House.

Kohn, M. (2013). What is wrong with gentrification? *Urban Research and Practice*. Routledge, 6(3), pp. 297–310. doi:10.1080/17535069.2013.846006.

Lees, L. (2000). A reappraisal of gentrification: towards a 'geography of gentrification'. *Progress in Human Geography*, 24(3), pp. 389–408.

Lees, L. (2012). The geography of gentrification: Thinking through comparative urbanism. *Progress in Human Geography*, 36(2), pp. 155–171. doi:10.1177/0309132511412998.

Lees, L., Annunziata, S., and Rivas-Alonso, C. (2018). Resisting planetary gentrification: The value of survivability in the fight to stay put. *Annals of the American Association of Geographers*, 108(2). doi:10.1080/24694452.2017.1365587.

Lefebvre, H. (1991). *The production of space*. Oxford: Basil Blackwell. Available at: http://proxy.mah.se/login?url=http://search.ebscohost.com/login.aspx?direct=true&db=cat05074a&AN=malmo.b1001981&site=eds-live

LeGates, R. T. and Hartman, C. (1982). Gentrification-caused displacement. *Urban Law*, 14(1), pp. 31–55.

Ley, D. (1980). Liberal ideology and the postindustrial city. *Annals of the Association of American Geographers*. Association of American Geographers, Taylor & Francis, Ltd., 70(2), pp. 238–258. Available at: http://www.jstor.org/stable/2562952

Ley, D. and Yang, Q. (2017). Global gentrifications: Uneven development and displacement; and planetary gentrification. *The AAG Review of Books*, 5(2). doi:10.1080/2325548X.2017.1292585.

London, B. (1980). Gentrification as urban reinvasion: Some preliminarg definitional and theoretical consideration. In: Laska, S. B. and Spain, D. (eds.) *Back to the city: Issues in neighborhood renovation*. New York: Pergamon Press, pp. 77–92.

Maloutas, T. (2012). Contextual diversity in gentrification research. *Critical Sociology*, 38(1), pp. 33–48. doi:10.1177/0896920510380950.

Maloutas, T. (2018). Travelling concepts and universal particularisms: A reappraisal of gentrification's global reach. *European Urban and Regional Studies*, 25(3), pp. 250–265. doi:10.1177/0969776417709547.

Marcuse, P. (1985). Gentrification, abandonment, and displacement: connections, causes, and policy responses in New York City. *Washington University Journal of Urban and Contemporary Law*, 28, pp. 195–248.

Marcuse, P. (2010). A note from Peter Marcuse. *City*, 14(1), pp. 187–188.

McKinnish, T., Walsh, R., and White, K. T. (2010). Who gentrifies low-income neighborhoods? *Journal of Urban Economics*. Elsevier Inc., 67(2), pp. 180–193. doi:10.1016/j.jue.2009.08.003.

Meltzer, R. and Ghorbani, P. (2017). Does gentrification increase employment opportunities in low-income neighborhoods? *Regional Science and Urban Economics*, 66. doi:10.1016/j.regsciurbeco.2017.06.002.

Newman, K. and Wyly, E. (2006). The right to stay put, revisited: Gentrification and resistance to displacement in New York City. *Urban Studies*, 43(1), pp. 23–57. doi:10.1080/00420980500388710.

Porteous, J. D. and Smith, S. E. (2001). *Domicide: The global destruction of home, housing studies*. Montréal: MQUP. doi:10.1080/02673030320000087766.

Pull, E. and Richard, Å. (2019). Domicide: Displacement and dispossessions in Uppsala, Sweden. *Social and Cultural Geography*. Routledge, 00(00), pp. 1–20. doi:10.1080/14649365.2019.1601245.

Relph, E. C. (1976). *Place and placelessness*. Pion, Limited (Research in Planning and Design). Available at: https://books.google.se/books?id=XHyxAAAAIAAJ

Sassen, S. (2014). *Expulsions – Brutality and complexity in the global economy*. London: The Belknap Press of Harvard University Press.

Shin, H. B. (2018). Urban Movements and the Genealogy of Urban Rights Discourses: The Case of Urban Protesters against Redevelopment and Displacement in Seoul, South Korea, *Annals of the American Association of Geographers*, 108(2), pp. 356–369.

Slater, T. (2006). The eviction of criticial perspectives from gentrification research. *International Journal of Urban and Regional Research*, 30(4), pp. 737–757. doi:10.1111/j.1468-2427.2008.00772.x.

Smith, N. (1979). Toward a theory of gentrification: A back to the city movement by capital, not people. *Journal of the American Planning Association*, 45(4), pp. 538–548. doi:10.1080/01944367908977002.

Smith, N. (1982). Gentrification and uneven development. *Economic Geography*. Clark University, Wiley, 58(2), pp. 139–155. doi:10.2307/143793.

Smith, N. (2002). New globalism, new urbanism: Gentrification as global urban strategy. *Antipode*, 34(3), pp. 427–450. doi:10.1111/1467-8330.00249.

Thursz, D. (1966). *Where are they now? A study of the impact of relocation on former residents of southwest Washington who were served in an HWC demonstration project*. Washington DC: Health and Welfare Council of the National Capital Area.

Tuan, Y.-F. (1971). Geography, phenomenology, and the study of human nature. *Canadian Geographer*, xv(3), pp. 181–192.

Tuan, Y.-F. (1979). Space and place: Humanistic perspective. *Philosophy in Geography*, pp. 387–427. doi:10.1007/978-94-009-9394-5_19.
Valli, C. (2015). A sense of displacement: Long-time residents' feelings of displacement in gentrifying Bushwick, New York. *International Journal of Urban and Regional Research*, 39(6), pp. 1191–1208. doi:10.1111/1468-2427.12340.
Vigdor, J. L. (2002). Does gentrification harm the poor? *Brookings-Wharton Papers on Urban Affairs*, 2002(1), pp. 133–182. doi:10.1353/urb.2002.0012.
Vigdor, J. L. (2010). Is urban decay bad? Is urban revitalization bad too? *Journal of Urban Economics*. Elsevier Inc., 68(3), pp. 277–289. doi:10.1016/j.jue.2010.05.003.
Wacquant, L. (2008). Relocating gentrification: The working class, science and the state in recent urban research. *International Journal of Urban and Regional Research*. Blackwell Publishing Ltd, 32(1), pp. 198–205. doi:10.1111/j.1468–2427.2008.00774.x.
Wyly, E. (2015). Gentrification on the planetary urban frontier: The evolution of Turner's noösphere. *Annual Lecture Article Urban Studies Urban Studies Journal Limited*, 52(14). doi:10.1177/0042098015601362.
Zuk, M., Bierbaum, A.H., Chapple, K., Gorska, K., Loukaitou Sideris, A., Ong, P., and Thomas, T. (2015). *Gentrification, Displacement and the Role of Public Investment: A Literature Review*. Berkeley: University of California.

1 Council estate renewal in London

The challenges of evidencing its gentrification-induced displacement

Phil Hubbard and Loretta Lees

Introduction

It was in London that Ruth Glass famously first diagnosed (classic) gentrification in the 1960s, alerting us to the class-based improvement of inner-city districts that, over time, displaced low-income residents and inverted many of the key assumptions of the Chicago School's ecological models dominant at the time (see Lees et al., 2008, Chapter 2). But this 1960s gentrification was by no means the first wave of activity displacing lower-income residents from the capital, with earlier rounds of state-sponsored redevelopment having targeted inner-city 'slums' in the immediate post-war era, decanting large numbers of lower-income residents to distant New Towns and satellite communities to enable the development of council estates built on modernist assumptions (Tunstall and Lowe, 2012). The scale of ambition was impressive, with mass production techniques used to produce subsidized council housing at an unprecedented rate and nearly half a million new homes added to London's housing stock by the 1960s. Young and Wilmott (1957) famously documented the problems that beset those displaced from the East End 'slums' as they struggled to acclimatize to their new council estate surroundings: anonymity, isolation, status competition, lack of facilities, and even neurosis. The mass removal of established communities – and a programme of dispersal that tore apart existing social networks – was to have long-lasting legacies, and the attempts to rebuild functioning communities were often fraught and drawn-out affairs.

More recently, from the late 1990s and into the twenty-first century, London's council estates have been facing a 'new' urban renewal that threatens to repeat many of the mistakes of post-war urban renewal, which displaced and disrupted local communities and exacerbated the social problems slum clearance was meant to solve. With the nation's highest land values and greatest housing pressures, London is at the forefront of a 'new' urban renewal that has identified many of its council estates as fit only for demolition; the rhetoric of 'sink estates' has come to the fore, with politicians bemoaning the concentrated poverty and social malaise that they identify as characteristic of post-war council estates. For example, the much-maligned Aylesbury estate in Southwark was the launching pad for New Labour's urban renaissance vision in the late 1990s, a vision that promoted mixed-income communities to replace estates that were depicted (often wrongly) as crime-ridden and dysfunctional (Lees, 2014a). More recently, the Coalition and Conservative governments have similarly spoken of regenerating neglected estates, and they have used the imagery of brutal modernist architectural estates constructed in the 1960s as a way of promoting a new wave of regeneration that will supposedly re/connect or knit these estates 'back to the mainstream' (Campkin, 2017).

While the rhetoric of sink estates (Slater, 2018) glosses over the diversity of community and spatial forms these estates take, the idea that something needs to be done about London's post-war council estates is now largely taken for granted. Analyses that identified the pivotal role estate residents played in the English riots of 2011 (see Till, 2013), as well as current concerns about knife crime in the capital, help consolidate the idea that these spaces need to be regenerated. At the same time, the housing affordability crisis is also being used to argue that many of the larger, high rise estates – once depicted as high density – are not densely-occupied enough. A government-sponsored report by Savills in collaboration with the Space Syntax group promotes the adoption of a 'complete streets' model of redevelopment rather than a 'block-based' one (Campkin, 2017; see Create Streets, http://dev.createstreets.com/). This type of model posits that high-rise development surrounded by communal and collective areas of public space (e.g., gardens, playgrounds, and drying areas) breeds neglect of public space, an argument that can be traced back to 'defensible space' arguments in the 1970s and 1980s (see Jacobs and Lees, 2015; Lees and Warwick, forthcoming); is wasteful of resource; and does not constitute 'the best and highest' form of land use. This implies that the redevelopment of London's council estates is morally, socially, and economically necessary – with estates themselves being identified as *de facto* brownfield sites and hence suitable for new housing even though they are already home to long-established communities, many of these multi-class, multi-cultural, and multi-ethnic in character.

Given current budgetary constraints, few local authorities are in a position to redevelop these estates alone; the preferred solution is to transfer these assets to private developers, who will demolish most of what currently stands and replace it with a mix of affordable and market rate housing, increasing the overall number of housing units available. While this means that state assets are being ceded to the private sector, the need to increase the overall housing supply via densification of these estates justifies this solution. The Mayor of London's Housing Strategy (2014: 59) called for the 'vast development potential in London's existing affordable housing estates' to be unlocked through private redevelopment. To 'kick-start and accelerate' that process, the Tory government launched a GBP 150 million Estate

Regeneration Programme of loans to private developers 'redeveloping existing estates' on 'a mixed tenure basis' (Homes and Communities Agency, 2014), continuing New Labour's mixed community policy of 'gentrification by stealth' (Lees, 2008; Bridge et al., 2011). The London Assembly (2015: 14) estimated that in the preceding decade, fifty former council estates across London received planning permission for partial or complete demolition and redevelopment at higher densities. As we describe below, this appears a gross under-estimate of the extent of the demolitions undertaken; irrespective, the number of households 'decanted' from these estates is clearly considerable, and the cumulative impacts on residents (both tenants and leaseholders) who have already been displaced or are still waiting to be decanted is potentially life-changing.

The displacements associated with the redevelopment of London's council estates are significant in number and escalating, but as yet, there has been little attention given to these – most of the extant literature on this regeneration focuses on resistance rather than the rehousing of those who are removed on either a short or long-term basis (see, e.g., Watt, 2016). There are manifold reasons for this, not least the fact that 'displacement is much harder to detect than gentrification' (Chapple and Loukaitou-Sideris, 2019). As we describe below, while it is theoretically possible to use existing sources to demonstrate the socio-economic uplift of specific neighbourhoods, the range of data available for actually tracking the journeys of the displaced is frustratingly limited. Governmental agencies are seemingly uninterested in where former council estate residents move to, only concerned that they have been 'decanted' to allow demolition to begin. This means we are left to piece together disparate evidence to try to estimate the extent of displacement associated with the redevelopment of London's housing estates – something that matters profoundly given the potential impact that rehousing has on the thousands affected by this 'new urban renewal'.

The challenge of evidencing displacement from London's council estates

The focus on displacement (and those displaced) has a patchy history in the gentrification literature, with recent commentary lamenting the effective '"displacement" of displacement' (Helbrecht, 2018: 2). Work in gentrification studies has historically tended to focus on middle-class gentrifiers and the production of gentrified living spaces (Slater et al., 2004; Huse, 2014; Paton, 2014) rather than the consequences for low-income groups. Hence, Helbrecht (2018: 2) describes the gentrification literature as 'a one-eyed cyclops that operates with an enormous intellectual bias because it observes only the upgrading aspect of the gentrification process while ignoring displacement'. Displacement has consequently been coined as the 'dark side' of gentrification (Baeten et al., 2017: 645), an observation that begs a more detailed investigation of the different forms and modalities of gentrification-induced displacement.

In many ways, the lack of data on the numbers displaced has allowed governments, policy-makers, and planners to pursue strategies of gentrification unchallenged by statistical evidence of what is often mooted as its most negative impact: the displacement of long-term residents (Atkinson, 2000). This point is doubly relevant in the context of the state's effective hand-over of council estates in London to private developers. While local government appears to see displacement of tenants and leaseholders a price worth paying in return for an increased housing supply in the capital, if those displaced cannot afford (or find) decent housing elsewhere or cannot afford to return to the redeveloped estate once complete, then the costs of renewal may well outweigh the benefits for the city's population. Here, the social costs of displacement – namely, the psychological impact of moving to a new neighbourhood and potentially failing to re-establish meaningful social networks and connections – are potentially massive if one considers the 'decanting' of residents as a form of violence.

Foregrounding violence in discussions of displacement may seem extreme when, in some studies, the effects of displacement seem relatively benign (but see Elliot-Cooper et al., 2019). For instance, Young and Wilmott's (1957) classic study of kinship in east London identified many individuals who actually found displacement to have a beneficial impact on their lives, with an enforced move from the inner city to their new estates bringing them heating, running water, indoor toilets, and multiple bedrooms. Longitudinal research in Glasgow by Kearns and Mason (2013, 2015) likewise suggests that there might be a difference in the 'psychosocial' impacts of displacement between those willing to move and those who are reluctant displacees. Reporting deleterious health outcomes for those displaced from central Glasgow housing estates, their conclusion was that 'most of those who moved considered that they had "bettered" their residential conditions, though again less so in neighbourhood than in dwelling terms' (Kearns and Mason, 2013: 195). The latter observation is important given the argument that 'working-class' people are said to exhibit a phenomenological understanding of their home and neighbourhood as a 'comfortable lived space' rather than a financial investment (Davidson and Lees, 2010). Thus, even if displaced residents receive the market value for their loss of property, it is impossible to compensate them for the longing and isolation that are often felt when their home is lost. In some cases, a new place may never feel truly like home; no matter how many new friends are made or how much better a new house may be, the memories of their original home and neighbourhood will always remind the displaced of their loss (Jones, 2015). The paradox here is that the 'objective' social good which derives from moving to a 'better' neighbourhood becomes a form of 'systemic violence' – not always a physical violence directly executed by individuals, but one that 'operates anonymously, systemically and invisibly through the very way society is organised' (Baeten et al., 2017: 643).

Of course, much here depends on where displaced residents relocate to. Crawford and Sainsbury (2017) argue that rehousing displaced residents across a range of locations may contribute to a loss of social networks and associated social capital (see also Posthumus et al., 2013). Lyons (1996) reports that, given the choice, lower-status households tend to move more locally than more affluent ones, reflecting both their restricted choices as well as their desire to maintain localized social networks. Atkinson (2003) suggests that this represents a somewhat 'desperate' attempt by residents to maintain a foothold near the locations they have come from. However, where displacees relocate to ultimately has significant consequences in terms of their ability to construct meaningful social ties. Several US studies conclude that there is little successful integration of displaced households into more distanced communities (Goetz, 2003; Newman and Wyly, 2006; Greenbaum et al., 2008). Of course, this is a generalization, and it has been noted that younger residents find it easier to adapt than older ones. Those who have lived longest in their original community appear to gain fewest benefits from relocation (Van Criekingen, 2008). Indeed, older residents are usually reluctant to engage with medical services in their new neighbourhood, and they sometimes travel long distances to engage with GPs and pharmacists they are familiar with (Crawford and Sainsbury, 2017). Kleinhans (2003) suggests that, in addition to age, 'personality' can be important in shaping experiences of displacement, with more resilient individuals able to take a more positive view of the 'relocation' process.

Although there is now emerging evidence of the negative impacts of gentrification-induced displacement, to date there has been little examination of the impacts of the redevelopment of London's council estates on those displaced. Therefore, we began our own investigation with the aim of trying to *measure* the extent of displacement and *track* where the council residents were being displaced to. Here, the first challenge was identifying the number of redevelopment schemes involving council estates in London since 1997, when New Labour's initial identification of the need for urban renewal began. While one would imagine that there is a central database of council estates in London, the situation is complicated for several reasons: There are thirty-two London boroughs with their own council housing estates, and although it is possible to get information on their current holdings, many of these estates have been ceded to the private sector through stock transfer and no longer appear as council housing per se; indeed, they are now typically operated by registered social landlords or housing associations (Watt, 2009). There is also the thorny question of what is meant by an estate, as not all council housing in London sits on the types of contiguous sites that fit into a common-sense definition of an 'estate' – the latter defined by the provision of certain shared amenities as well as a distinct 'sense of place'. Many 'estates' are actually low-rise houses that are relatively dispersed across a number of streets; in contrast, the estate redevelopment programme has tended to focus on estates which are a mix of high-rise blocks, maisonettes, and houses that are morphologically distinct and managed as a single entity.

Freedom of information requests, analysis of planning databases, and trawls of websites led us to ultimately identify 161 estates in London of more than 100 households where there has been a scheme to demolish and redevelop housing since 1997. This excludes estates that had been stock transferred before 1997 as well as those that have always been run by housing associations, meaning some schemes identified by others were not included in our database. Nonetheless, we estimated that 190 schemes on these 161 sites entailed the demolition of at least 55,000 homes, with the 'average' scheme involving the decanting of 274 households prior to demolition and rebuild. Some schemes were relatively modest in scale. For example, the Triangle estate in Islington involved the demolition of just six maisonette dwellings, a garage, and one retail unit to enable the construction of fifty-four new infill homes, half of which were available for affordable rent. In this case, one residential leaseholder and one commercial leaseholder were served with Compulsory Purchase Orders by the local council. On the contrary, some schemes have involved mass demolitions and drawn-out decanting programmes whereby council tenants were moved elsewhere and leaseholders bought out. For instance, the Heygate estate in Southwark involved the demolition of 1,200 units, the South Acton estate in Ealing 1,998 units, and the Woodberry Down estate in Hackney 1,980 units. In all, twelve schemes had more than 1,000 units demolished, suggesting that in some cases large and established communities were being broken up by processes of displacement.

Nevertheless, even knowing how many households might have been affected does not establish how many people have been impacted. According to the 2011 census, the average household renting from a council consisted of 2.4 people – suggesting around 131,000 people have been displaced, as a rough estimate. Further, we have no idea where these people have moved to, but a couple of studies give us some clues. Davis and Thornley (2010) examined the compulsory purchase of the Clays Lane estate (Newham) before the London 2012 Olympics, where 424 tenants were decanted to allow for the development of the Olympic site. The majority of these tenants were relocated to East London's outer boroughs (Waltham Forest and Barking & Dagenham), with only a handful being rehoused in the same borough. In Lees's work with local campaigners (London Tenants Federation et al., 2014), the displacees from the Heygate estate in Southwark were found through local contacts on the ground. As their maps show, the council tenants were displaced from their SE17 postcode: some managed to stay in Southwark, some ended up in Lambeth, and some were rehoused in more distant parts of London. Leaseholders who were bought out ended up in a much more dispersed pattern, suggesting the sums they received were not sufficient to allow them to resettle in inner London: some relocated to parts of Kent, Essex, and Berkshire, as well as outer London boroughs (see maps on pages 8–9 in London Tenants Federation et al., 2014).

Piecing together the extent of displacement from London's council estates is a task that relies on combining data that indicates when households left a given estate as well as the locations they relocated to. Doing this for all 161 estates that have undergone some decanting is a painstaking process, and we anticipated that instead of multiple individual freedom of information requests, we would use the type of secondary sources routinely employed in studies of gentrification-induced displacement. The requirements (criteria) of the data needed to be as follows:

- The data needed be longitudinal (measured over time) to enable the tracking of displacement from one time point to another.
- The data needed to cover as much of the last twenty years as possible (i.e., from 1997 to 2017 the study period selected for the research).
- The data needed to be available at a small area level (e.g., output area or postcode) so that data could be located as near as possible to estate-level.
- The data should not have contained large gaps in time, noting people could have moved multiple times in the interim.

Most quantitative studies of gentrification-induced displacement (see Easton et al., 2019, for an extended review of studies in the United Kingdom, North America, and beyond) have employed national censuses and/or local survey data – data that come with significant limitations in terms of revealing actual patterns of intra-urban migration and displacement. The UK census, for example, provides a snapshot of the social-economic make-up of small-scale neighbourhoods every ten years, allowing some inferences to be drawn about the extent to which low-income residents have been replaced by more affluent ones (see Figure 1.1). Invariably, the utility of such periodic data collection for inferring displacement depends not only on the currency of the data, but also the salience of the questions asked: for example, the omission of data relating to household income, rent, or house prices in the UK census has long-frustrated attempts to explore housing affordability, and the lack of questions about someone's previous address and where they moved from makes it hard to distinguish between displacement and the replacement that Hamnett (2003) has claimed for London.

The problems of tracking migration in and out of London estates using published census sources led us to apply for census 'flow' data. This data gave an origin and destination output area for all respondents included in the 2011 census who lived somewhere different in 2010. We applied for this secure dataset, which is broken down by tenure (social rented/private-rented/owner-occupied) at the level of output area. Using the list of schemes we had previously identified, we were able to map migration flows for social tenants and identified outflows from several renewal estates where the timings of the

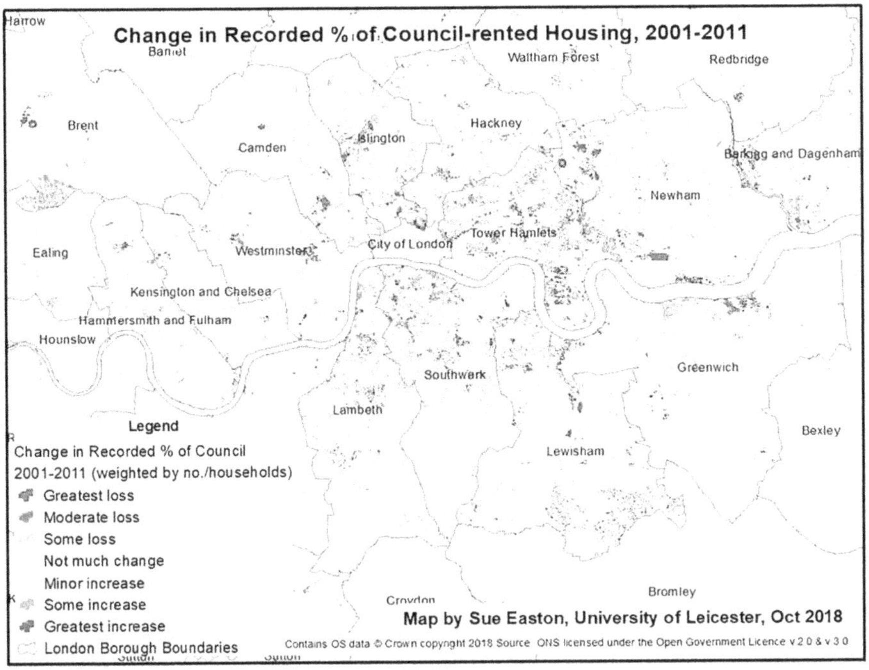

Figure 1.1 Change in recorded % of council-rented housing, 2001–2011.
Source: Sue Easton, University of Leicester, Oct, 2018.

demolition fitted. This delivered our best evidence to date (see Figure 1.2), but the disadvantage was that the data was only available for a single year, 2010–2011.

Another key dataset that we proposed for analysis as part of the project was GP Patient Registration (GPPR) data from GP/health centres. In theory, this data would have located potential council estate residents and shown the migration of the key population groups we wanted to track. However, after completing a laborious application, we were not given access. Therefore, we applied for other sources of data, such as National Pupil Data from the Department for Education. With this data, we undertook a pilot analysis of outmigration flows for nine large 'test' estates that we thought might show us evidence of the displacement of families with children. However, significant outflow was only observed for one of the estates – the Heygate estate in Southwark, between 2005 and 2008 (see Table 1.1).

We also sought permission to analyse the secure version of the Continuous Record of Social Housing (Ministry of Communities, Housing and Local Government) for 2007–2017. This dataset contains administrative data

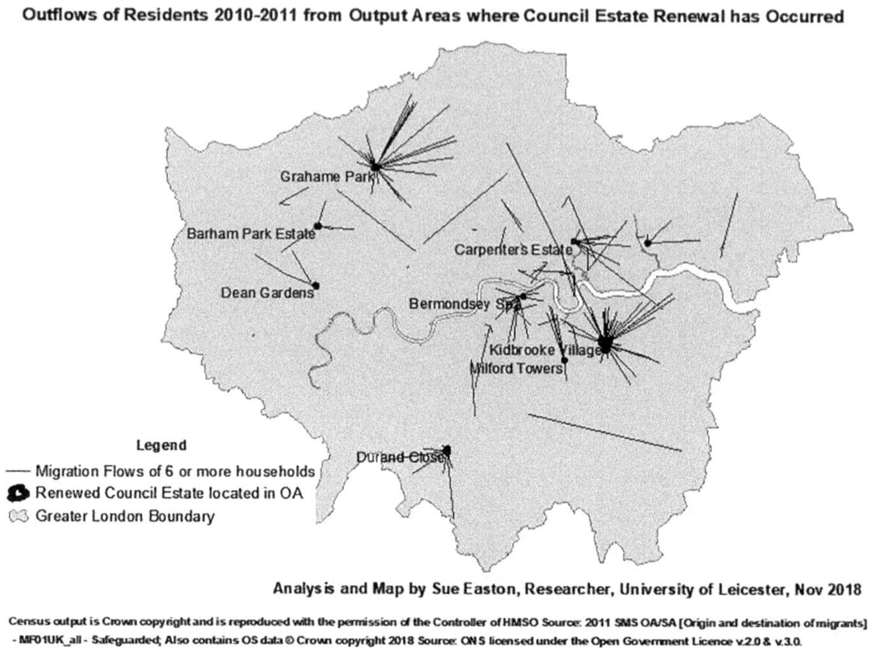

Figure 1.2 Outflows of residents, 2010–2011.
Source: Sue Easton, Nov, 2018.

Table 1.1 Outflow of Heygate estate

Heygate Estate			
Moved?			
Year	*No*	*Yes*	*Total*
2002	31	20	51
2005	30	29	59
2008	–	67	–
2011	0	0	0
2014	0	0	0
Total:	61	116	–

Note: – denotes suppressed due to statistical disclosure control.
Data Source: From National Pupil Database extracts *provided under licence from the Department for Education.* Analysis by Sue Easton, Research Associate, July 2017.

from local authorities on social lettings, evictions, reason for rehousing in council housing, and so on for both council and social (housing association) managed homes. Again, we detected only minimal migration flows among council tenants. In addition, we applied for various other datasets, such as Electoral Roll and Consumer Register data covering 1998–2017 from the Consumer Data Research Centre, which can be used to produce local estimates of population turnover at lower super output area level. However, upon analysis, we found the degree of year-on-year variability in the data too great to support robust longitudinal analysis.

In frustration over the various issues with official data and the fact that we were not getting the quantitative evidence we desperately desired, in late summer 2018, we decided to request data on the displacement of council tenants, leaseholders, and homeless households in temporary accommodation outside of London directly from London Borough Councils via Freedom of Information requests. This was undertaken for all thirty-three local authorities, including the City of London Corporation. However, as we expected, there was considerable variability in the type and quality of the data returned, which prevented comparability of most of the data. A key obstacle was that local authorities did not seem to be able to distinguish whether households had been placed 'out-of-area' due to persecution (e.g., fleeing domestic violence) or other reasons, including estate renewal. Thus, it was not possible to identify the precise level of displacement resulting from estate renewal as opposed to domestic violence, anti-social behaviour, or other. The frustration here is, of course, that councils should have this information (e.g., through council tax change of address data) and be able to reveal it when subject to Freedom of Information requests. However, councils clearly have little interest in collating this information for estates themselves – especially in situations where this information demonstrates that large numbers of tenants being decanted from council or former council estates are now being rehoused outside the borough.

The frustrations of trying to estimate the extent and form of displacement from London's council estates are manifold. However, in a displacement context, access to viable sources of data that enable the tracking of individuals through space and across time is most important. Until such data are collected or made available, the extent of residential gentrification-induced displacement will remain largely unrecorded and invisible. In the meantime, we appear reliant on proxies for actual displacement, such as broad indicators of population churn, changes in owner occupation, or changes in the ethnic and class make-up of particular neighbourhoods (e.g., as revealed in census profiles). While such measures can be suggestive of the occurring involuntary displacement, they are rarely conclusive (see Easton et al., 2019). Rather than being measures of displacement per se, these are perhaps best thought of as measures of *displacement pressure* (Marcuse, 1986).

Conclusions

Our failure to reveal actual flows of displaced people from council estates being redeveloped in London suggests that, for the moment, we must rely on a mixture of proxy measures, approximations, and predictions that reveal tendencies but that cannot be relied upon to distinguish between involuntary displacement from council estates, voluntary outmigration, or incumbent socio-economic uplift. Part of the problem here is working out when a given scheme for redevelopment begins. In many cases, plans for regeneration are mooted years before any formal decanting or demolition begins, meaning that displacement actually starts much earlier than any formal decision or ballot about estate redevelopment. Accordingly, leaseholders and others who are more socially mobile may be looking to move from the estate long before they are required to leave. Tenants might also seek re-housing through a variety of means, not wanting to wait until they are formally decanted (see, e.g., Thompson et al., 2017, on the way tenants can use mental health status as a means to acquire better housing elsewhere in estates awaiting redevelopment). In many estates awaiting development, it is apparent that councils have enacted a gradual decline in services and facilities; this is known as 'managed decline', and it pushes people out. In estates that have been fully decanted bar a few people refusing to move, London councils have even enacted a more aggressive 'state Rachmanism' designed to force those residents out; for instance, on the Heygate estate, the council turned off their gas and electricity, mail was no longer delivered, and bailiffs literally carried residents out in the end (see Lees, 2014b).

If we accept the premise that displacement can be a form of violence (cf. Elliot-Cooper et al., 2019), the implication is that we need to be watchful for both direct, short-term displacements and longer-term, indirect ones and not leap to conclusions about whether displacement is causing harm on the basis of one-time snapshots of change (e.g., migration data that looks at changes over a single year). While the decanting of residents from London's council estates appears to be a singular act, enacted, and enforced by local authorities pursuing Compulsory Purchase Orders and removing council tenants, it can also occur through a series of smaller aggressions that displace industries and businesses, as well as residents, over a much longer time span. For example, the Carpenter's estate in Newham, East London, was first ear marked for renewal in 2000, and residents have seen (off) numerous plans since then. Most recently, the estate was earmarked for redevelopment in 2017, with the council issuing a tender to find a commercial partner to start redeveloping it in 2020. But this search was abandoned in November 2018, when a major reconsideration of the plans for the estate was mooted. In the meantime, some residents in James Riley Point, one of the three tower blocks on the estate, have had to contend with flooding issues as water poured down through their roof and workmen struggled to access flats vacated by tenants when the block was infested with ants some years earlier;

while some tenants were able to argue living with the ant infestation was intolerable, those who had bought their flats under the 'right to buy' scheme remained in situ, awaiting for news about what they might be offered for their flats. The potential psychological and physical consequences of living in this abeyance are multiple: the tortuous and exhausting processes of establishing how displacement will impact on one's home-space lead to feelings of shame, stress, and anxiety (Wallace, 2015). This can ultimately wear down individuals, leading to an inertia that makes effective resistance to displacement impossible (Lancione, 2017) or to necessary practices of 'survivability' that are themselves a form of resistance (see Lees et al., 2018).

The displacement associated with the renewal of London's council estates is rarely the product of a one-off event. More often it is a form of slow, percolating violence. Displacement is not just about direct replacement of poorer groups by wealthier ones; it also involves forms of social, economic, and cultural transition that alienate established populations. This can entail forms of slow violence that render particular neighbourhoods less hospitable and accommodating to established residents over time, as well as direct and forceful acts of expropriation which the vulnerable and precarious seem least able to cope with. In this regard, in addition to quantitatively tracking displacement itself, it appears vital that we evolve better qualitative measures of *displacement pressures* in order to predict, and ideally help prevent, future displacements.

London is at the forefront of a 'new' urban renewal that deems council estates as viable sites for redevelopment, where 'affordable' (not council) housing can be interspersed with new market-rate housing sold on for profit. From the developers' perspective, this is attractive because it represents an opportunity to make large profits by selling on 'luxury' flats and homes whilst fulfilling a social obligation by selling affordable, intermediate, and shared housing when it is in high demand. From the local authorities' point of view, such schemes enable them to deliver new 'affordable' housing whilst updating and renewing housing estates viewed in need of investment they cannot afford under current budgetary constraints. From the perspective of central government, such schemes are attractive in so much as they deliver urban regeneration by introducing more affluent residents to such estates, allegedly reducing deprivation through social mixing (Lees, 2008, 2014a). All of these logics can certainly be questioned, not least the idea that estate densification delivers more affordable housing. On the 161 estates we have examined, redevelopment will be adding around 45,000 additional units; however, the majority of these (fifty-two per cent) are sold at a market rate, and many schemes still fall well short of the thirty-five per cent affordable target set by London Mayor Sadiq Khan in 2016. Moreover, residents wanting to move back into the redeveloped estates as social (not council) housing tenants will often face inflated service charges, rents, and council tax. Much of the 'affordable' housing developed is not genuinely affordable (affordable housing is categorised as eighty per cent of market rents), and the new 'social housing' (around twenty-five per cent of the total) is not council housing (the cheapest form of housing in the United Kingdom).

We have to ask the following question: is this worth the disruption and displacement that the redevelopment of these estates necessitates? Given that over 50,000 families and households have seen their home demolished and experienced anxiety and uncertainty, sometimes over a prolonged period, all for the sake of around 7,000 additional affordable homes for Londoners, the answer appears to be negative. Nevertheless, it is hard to be definitive about this given the data deficit that surrounds estate redevelopment. To challenge current orthodoxies concerning the 'sacrifice' of existing communities for the sake of additional housing in the capital, we not only need further qualitative evidence of the toll this has taken on some households, we also need a better understanding of where the displaced have relocated to – or been relocated to. Thus, while some urban scholars have stridently challenged the very idea of estate renewal as 'gentrification by stealth' (Bridge et al., 2011), intended to privatize social housing and socially cleanse the inner city of low-income communities, the lack of empirical evidence on the negative impact of displacement means effective opposition remains difficult.

References

Atkinson, R. (2000). The hidden costs of gentrification: Displacement in Central London. *Journal of Housing and the Built Environment*, 15, pp. 307–326.

Atkinson, R. (2003). Introduction: Misunderstood saviour or vengeful wrecker? The many meanings and problems of gentrification. *Urban Studies*, 40(12), pp. 2343–2350.

Baeten, G., Westin, S., Pull, E., and Molina, I. (2017). Pressure and violence: Housing renovation and displacement in Sweden. *Environment and Planning A*, 49(3), pp. 631–651.

Bridge, G., Butler, T., and Lees, L. (eds.). (2011). *Mixed communities: Gentrification by stealth?* Bristol: Policy Press.

Campkin, B. (2017). Out-of-sync estates. In: Guillery, P. and Kroll, D. (eds.) *Mobilising housing histories: Learning from London's past*. London: RIBA Publishing, pp. 163–176.

Chapple, K. and Loukaitou-Sideris, A. (2019). *Transit-orientated displacement or community dividend? Understanding the effects of smart growth on communities.* Cambridge, MA: MIT Press.

Crawford, B. and Sainsbury, P. (2017). Opportunity or loss? Health impacts of estate renewal and the relocation of public housing residents. *Urban Policy and Research*, 35, pp. 137–149.

Davidson, M. and Lees, L. (2010). New-build gentrification: Its histories, trajectories, and critical geographies. *Population, Space and Place*, 16, pp. 395–411.

Davis, J. and Thornley, A. (2010). Urban regeneration for the London 2012 Olympics: Issues of land acquisition and legacy. *City, Culture and Society*, 1, pp. 89–98.

Easton, S., Lees, L., Hubbard, P., and Tate, N. (2019). Measuring and mapping displacement: The problem of quantification in the battle against gentrification. *Urban Studies*, 57(2), pp. 286–306.

Elliot-Cooper, A., Hubbard, P., and Lees, L. (2019). Moving beyond Marcuse: Gentrification, displacement, and the violence of un-homing. *Progress in Human Geography*. doi: 10.1177/0309132519830511.
Goetz, E. G. (2003). *Clearing the way: Deconcentrating the poor in urban America*. Washington, DC: The Urban Institute.
Greenbaum, S., Hathaway, W., Rodriguez, C., Spalding, A., and Ward, B. (2008). Deconcentration and social capital: Contradictions of a poverty alleviation policy. *Journal of Poverty*, 12, pp. 201–228.
Hamnett, C. (2003). Gentrification and the middle-class remaking of inner London, 1961–2001. *Urban Studies*, 40, pp. 2401–2426.
Helbrecht, I. (2018). Gentrification and displacement. In: Helbrecht, I. (ed.) *Gentrification and resistance: Research displacement processes and adaption strategies*. Wiesbaden: Springer, pp. 1–8.
Homes and Communities Agency. (2014). *Estate regeneration programme: Prospectus for 2015/16*. London: DCLG.
Huse, T. (2014). *Everyday life in the gentrifying city: On displacement, ethnic privilege and the right to stay put*. London: Routledge.
Jacobs, J. and Lees, L. (2013). Defensible space on the move: Revisiting the urban geography of Alice Coleman. *International Journal of Urban and Regional Research*, 37(5), pp. 1559–1583.
Jones, O. (2015). 'Not promising a landfall …': An autotopographical account of loss of place, memory and landscape. *Environmental Humanities*, 6, pp. 1–27.
Kearns, A. and Mason, P. (2013). Defining and measuring displacement: Is relocation from restructured neighbourhoods always unwelcome and disruptive? *Housing Studies*, 28, pp. 177–204.
Kearns, A. and Mason, P. (2015). Regeneration, relocation and health behaviours in deprived communities. *Health and Place*, 32, pp. 43–58.
Kleinhans, R. (2003). Displaced but still moving upwards in the housing career? Implications of forced residential relocation in the Netherlands. *Housing Studies*, 18, pp. 473–499.
Lancione, M. (2017). Revitalising the uncanny: Challenging inertia in the struggle against forced evictions. *Environment and Planning D: Society and Space*, 35(6), pp. 1012–1032.
Lees, L. (2008). Gentrification and social mixing: Towards an urban renaissance? *Urban Studies*, 45(12), pp. 2449–2470.
Lees, L. (2014a). The urban injustices of New Labour's 'new urban renewal': The case of the Aylesbury Estate in London. *Antipode*, 46, pp. 921–947.
Lees, L. (2014b). The death of sustainable communities in London? In: Imrie, R. and Lees, L. (eds.) *Sustainable London? The future of a global city*. Bristol: Policy Press, pp. 149–172.
Lees, L. and Warwick, E. (2021). *Defensible space: Mobilisation in English housing policy and practice*. RGS-IBG Book Series, Wiley.
Lees, L., Annunziata, S., and Rivas-Alonso, C. (2018). Resisting planetary gentrification: The value of survivability in the fight to stay put. *Annals of the Association of American Geographers*, 108(2), pp. 346–355.
Lees, L., Slater, T., and Wyly, E. (2008). *Gentrification*. New York: Routledge.
London Assembly. (2015). *Knock it down or do it up? The challenge of estates regeneration* (February). London: GLA.
London Tenants Federation, Lees, L., Just Space, and SNAG. (2014). *Staying put: An anti-gentrification handbook for council estates in London*. London: Calverts

Co-operative. Available at: https://justspace.org.uk/2014/06/19/staying-put-an-anti-gentrification-handbook-for-council-estates-in-london/

Lyons, M. (1996). Gentrification, socioeconomic change, and the geography of displacement. *Journal of Urban Affairs*, 18, pp. 39–62.

Marcuse, P. (1986). Abandonment, gentrification, and displacement: The linkages in New York City. In: Smith, N. and Williams, P. (eds.) *Gentrification of the city*. Boston, MA: Allen and Unwin, pp. 121–152.

Mayor of London. (2014). *London housing strategy*. London: GLA.

Newman, K. and Wyly, E. K. (2006). The right to stay put, revisited: Gentrification and resistance to displacement in New York City. *Urban Studies*, 3, pp. 23–57.

Paton, K. (2014). *Gentrification: A working class perspective*. Chichester: Ashgate.

Posthumus, H., Bolt, G., and van Kempen, R. (2013). Why do displaced residents move to socioeconomically disadvantaged neighbourhoods? *Housing Studies*, 28, pp. 272–293.

Slater, T. (2018). The invention of the 'sink estate': Consequential categorization and the UK housing crisis. *The Sociological Review*, 66(4), pp. 877–897.

Slater, T., Curran, W., and Lees, L. (2004). Gentrification research: New directions and critical scholarship. *Environment and Planning A*, 36, pp. 1141–1150.

Thompson, C., Lewis, D. J., Greenhalgh, T., Smith, N. R., Fahy, A. E., and Cummins, S. (2017). 'I don't know how I'm still standing' a Bakhtinian analysis of social housing and health narratives in East London. *Social Science and Medicine*, 177, pp. 27–34.

Till, J. (2013). The broken middle: The space of the London riots. *Cities*, 34, pp. 71–74.

Tunstall, B. and Lowe, S. G. (2012). *Connected communities: 'Breaking up communities'?: The social impact of housing demolition in the late twentieth century*. Connected Communities discussion paper: White Rose Centre.

Van Criekingen, M. (2008). Towards a geography of displacement. Moving out of Brussels' gentrifying neighbourhoods. *Journal of Housing and the Built Environment*, 23(3), pp. 199–213.

Wallace, A. (2015). Gentrification interrupted in Salford, UK: From New Deal to 'Limbo-Land' in a contemporary urban periphery. *Antipode*, 47, pp. 517–538.

Watt, P. (2009). Housing stock transfers, regeneration and state-led gentrification in London. *Urban Policy and Research*, 27, pp. 229–242.

Watt, P. (2016). A nomadic war machine in the metropolis: En/countering London's 21st-century housing crisis with Focus E15. *City*, 20, pp. 297–320.

Young, M. and Willmott, P. (1957). *Family and kinship in East London*. London: Routledge and Kegan Paul.

2 Overcoming the limits of theory

Conceptual musings on displacement

Ipsita Chatterjee

Introduction

> People move all the time. The poor will have to move; without moving, the project is not feasible. Sixty per cent of the families to be displaced are migrants from elsewhere. If they could move to this place [banks of the Sabarmati] from towns and villages hundreds of kilometres away, they can move a few more kilometres.
>
> (Mohammad Pathan Aliyar Khan, 2006)

Mohammad Pathan Aliyar Khan (MPK) related to the above statement without batting an eye lid. He was among the families slated for displacement. MPK was also the leader of the Sabarmati Nagrik Adhikar Maanch (SNAM), which was the largest people's resistance movement against displacement along the banks of the Sabarmati river. SNAM had by then negotiated a stay-order from the local courts against Sabarmati Riverfront Corporation Limited (SRFDCL), which was organizing displacement and urban revitalization on behalf of the Ahmedabad Municipal Corporation in Ahmedabad city, India. At that time, inspired by Lefebvre's (1970) *La révolution urbaine*, I had been frequenting the Sabarmati in search of an urban resistance of the poor. I had learned from Lefebvre,

> Disputes reveal the ideologies [sic] characteristic of the groups or classes involved, including the ideology (or ideologies) of those who contribute to the development of projects, *ideological urbanism*. The intervention of "disputants" introduces conflict into social logics (socio-logic as ideo-logic). The possibility of dispute causes these logics to manifest themselves as ideologies and promotes confrontation, which is a measure of the degree of urban democracy. Urban revolution and concrete (developed) democracy coincide. The urban practice of groups and classes—that is, their way of life, their morphology—can only confront urban ideology in this way. And, in this way, disputes evolve into demands.
>
> (Lefebvre 1970: 136–137, emphasis in the original)

If disputes indeed reveal the ideological characteristics of groups and classes and if elitist urban renewal projects like the Sabarmati Riverfront Development Project (SRFD) represent the elitist ideological-morphology, then how is it that the largest resistance movement there does not confront the very heart of that ideology, that is, displacement? What is the SNAM's 'way of life' (Lefebvre 1970: 137), their slum morphology-ideology that does not manifest as what Lefebvre calls 'concrete democracy' and 'urban revolution'? What kind of a *resistance* is this that willingly assimilates with ideological urbanism without disputing its *ideological* basis? I was deeply struck by the tone of inevitability in the resistance organizer's statement. Even after I had written exhaustively on SRFD, done ethnographies of displacement and the resettlement process, the sense of inevitability in MPK's voice always haunted me. This article is an effort at conceptualizing displacement so as to probe at that sense of inevitability.

In the Global South, development-induced displacement has been the normal order of doing business. Any resistance to this has been framed as a struggle between modernity and an irrational, selfish impulse to hold on to non-existent idyllic village communities. MPK's statement should be viewed in the context of that 'normal order of doing business'. The ontological warfare between holding on and allowing development to proceed is often waged on the streets and in the media. The difficult dialogues often swiftly degenerate into 'Do you not want progress?' 'Do you not want development?' 'Ask not what the nation can do for you, ask what you can do for the nation?' In the context of the World Bank-funded Sardar Sarovar Dam project in India and people's resistance to it in the form of Narmada Bachao Andolan (Save Narmada Project), Roy (1999) explains how the entire talk surrounding displacement is obscured within an overarching framework of development versus anti-development. She cautions that whether we hate the dam or love it, we must be aware that a price is being paid for it. She also alerts us that usually those who are pro-displacement are not the ones being displaced, and those who are being displaced are usually against displacement and are immediately labelled as anti-development! Cernea (2000) contends that the last twenty years have seen the displacement and resettlement of some 200 million people. Despite many precautions on the part of the state, relocation after displacement did not prove to be a 'happily ever after' scenario. This seemed to be consistent for large infrastructure projects, like dams in both the Global South and the Global North. Koenig (2006) mentions the Tennessee Valley Authority in the United States, Mexico's Papaloapan dam, and India's Hirakud dam as examples. Parasuraman (2016) couches development-induced displacement as an extension of the colonial rationality – acknowledging, however, that use of land has always been defined by dominant groups (colonial or otherwise). In the context of colonial history, land grab acquires legal legitimacy (though various land acquisition acts) and is signified as the rational path of progress. Eminent domain or some version of it exists in all countries of the world, and whether it emerged in the colonial, racial, or class contexts, it continues to legitimize displacement – albeit, sometimes, with what is referred to as the 'just compensation'.

It is in the context of this larger framework of development-induced displacement, dam versus village, colonialism versus post-coloniality, displacement versus just compensation, that I situate MPK's statement. Growing up in India and going to college during the Save Narmada Campaign days, my consciousness has always been troubled by this unresolved dialectic between 'greater common good' and what the media characterized as a bunch of hysterical people who do not want electricity! MPK's home would be taken away and swept under malls, boulevards, and waterparks, yet in his unwavering calmness, he continued to rationalize that the poor have always moved and they will move again but, this time, with 'just compensation'. On the other hand, Clark (2009: 182) points out,

> Improperly implemented development-forced displacement can violate a multiplicity of human rights, leading in the worst cases to death, in the majority of cases to impoverishment, and in most cases to the negation of rights to adequate housing, security of tenure, food security, freedom of movement, and choice of residence.

MPK's adversaries argue that he has sold out people's resistance to the builders who now run the Ahmedabad Municipal Corporation (AMC) and is, therefore, happy to sing in tune with the powers that be. MPK's thesis confronts Lefebvre's argument that 'intervention of "disputants" introduces conflict' in 'socio-logics as ideo-logics'. Here, the disputants are not an easily identifiable homogenous cohort waiting for ethnographic representation of their concrete revolutionary urbanism. Ontologies of development, movement, migration, justice, land, rights, what is inevitable, and what is politically declared as inevitable are continuously being contested in the same city and same narrow stretch of the Sabarmati. What did I learn about displacement and resettlement from those contestations? How do they inform my conceptualization of displacement-resettlement in general? SRFD and the displacement-resettlement surrounding it informs my consciousness as I write this article, but I intend my conceptual musings to go beyond that case study.

Placing displacement?

In the Global South, displacement is the central conceptual lens through which development is understood and critiqued. This displacement is induced by large infrastructural projects like dams (Roy, 1999; Oliver-Smith, 2009; Nayak, 2013; van Cleef, 2016) or mines, ports, and irrigation projects (Parasuraman, 2016); through land acquisition acts and eminent domain (De Wet, 2006; Parasuraman, 2016); because of economic modernization and capitalism (Escobar, 2003; Haldar and Abraham, 2015); through state-induced urban revitalization (Carl LeVann and Olubowale, 2014); or because of neoliberal urban renewal (Bunnel et al., 2002; Marinelli, 2009; Chatterjee, 2014; Patel et al., 2015). The physical uprooting of the poor class, religious/ethnic minorities, and marginal groups from their homes and hearths and their forced resettlement elsewhere is understood as displacement. As the above literature shows, there is a general academic consensus that displacement is undesirable and constitutes injustice (Roy, 1999; Bunnel et al., 2002; Escobar, 2003; Marinelli, 2009; Oliver-Smith, 2009; Nayak, 2013; Carl LeVann and Olubowale, 2014; Chatterjee, 2014; Haldar and Abraham, 2015; Patel et al., 2015; Parasuraman, 2016; van Cleef, 2016). Some researchers critique state-sponsored (Carl LeVann and Olubowale, 2014) or capital-sponsored (Bunnel et al., 2002; Marinelli, 2009; Chatterjee, 2014; Patel et al., 2015) displacement, demanding a more socio-spatially just society. Others (De Wet, 2006; Parasuraman, 2016) point out the imperialistic moorings behind eminent domain and land acquisition acts. However, a quick peek into Google Scholar when combining the terms 'displacement' and 'Global South' reveals that an overwhelming proportion of the literature focuses on large-scale displacement ushered by mega infrastructural projects often involving resettlement across the affected state and sometimes even inter-state movements (e.g., Fisher, 1995; Roy, 1999; Oliver-Smith, 2009; Nayak, 2013; van Cleef, 2016). A smaller contingent of recent work explores intra-urban displacement due to urban revitalization, state-sponsored projects, or public-private partnerships (Bunnel et al., 2002; Bogaert, 2011; Chatterjee, 2014; Doshi, 2015; Patel et al., 2015), but this literature is very cautious about attributing causality to gentrification. Gentrification theory is arguably rooted in the Anglophone context, and therefore, researchers should avoid its cut and paste application on the cities of the Global South, which often have different socio-spatial trajectories of colonial, post-colonial, industrial and post-globalization socio-spatial realities. In fact, Chatterjee (2014) warns against simplistic theory-transfer from the Global North to the South, and Ghertner (2014) cautions that gentrification has Anglo-American roots and simple geographic diffusion to provincialize urban theory in order to make it more global may not work. Lees et al. (2015) argue that although gentrification strategies often serve as important tools for capital accumulation through housing development, we must be cautious as to how we understand displacement and urban renewal in the Global South. Roy (2011) argues that the megacity of the Global South represents a 'constitutive outside' to contemporary urban studies and calls for a subaltern urbanism that theorizes subaltern spaces and classes.

On the other hand, the literature on displacement from the Global North has a heavy but not exclusive emphasis on gentrification. For example, Atkinson (2004) argues that urban revitalization may induce displacement but may not involve class takeover of neighbourhoods that gentrification often entails – a claim that has been thoroughly debated (Marcuse, 1986;

Smith, 1996). On the other hand, Smith (1987: 462) introduces 'rent gap' as 'the gap between the actual *capitalized* ground rent (land value) of a plot of land given its present use and the *potential* ground rent that might be gleaned under a higher and better' use. The rent gap creates an opportunity for developers (Smith refers to them as 'pioneers' and 'urban cowboys') to restructure the city's core. This restructuring often manifests as urban renewal in the form of slum clearance and the clearance of obsolete industrial buildings and warehouses where the slum-dwellers worked. Marcuse (1986) emphasizes the displacement-inducing nature of gentrification, distinguishing between physical displacement that uproots residents because of housing stock regeneration and economic displacement through rise in rent. Other forms of displacement identified by Marcus (2013: 156) include 'exclusionary displacement', where a household can no longer enter their dwelling due to gentrification-related changes that affect that dwelling or its immediate surrounding. He also identifies 'displacement by abandonment', when public and private parties decide that investment to upkeep the neighbourhood is no longer warranted (157).

Based on a case study of London, Hamnett (2003: 182) claims that displacement is often 'secular replacement'. Secular replacement happens to neighbourhoods when people move out due to long term occupational and industrial change. Therefore, for Hamnett, gentrification can certainly cause direct or indirect displacement of the working class, but a lot of working-class homeowners also take advantage of the rise in property value to retire, sell out, and move to the suburbs. In fact, there is little evidence that gentrification is class revenge. Hamnett (2003: 183) argues that Smith's analogy of the urban pioneers (developers) colonizing the city and ousting its working class is rather 'melodramatic'. As a city transitions from industrial to post-industrial, the proportion of the industrial working class declines as a part of the 'normal' transition; it does not imply that the managerial class are 'squeezing manual workers out of gentrifying areas' (183). Similarly, Freeman and Braconi (2004) argue that empirical evidence does not clearly substantiate the displacement-inducing nature of gentrification; instead, they claim that low-income groups often stay behind to take advantage of better infrastructure, safer roads, and good retailing as the neighbourhood gentrifies. On the other hand, Slater (2006: 747) adopts a critical stance towards Hamnett (2003) and Freeman and Barconi (2004), claiming that they normalize oppressive processes like gentrification, which leads to the 'displacement of displacement' from gentrification literature. Slater (2006) laments that the methodological challenges of accumulating data on the displaced cause the erasure of research on displacement. Because data is hard to collect, displacement does not exist.

As I look through these debates and take stock of the available conceptual lenses sorted into Global South and Global North – sometimes docketed under 'development-induced', other times as 'gentrification-induced', sometimes explored by quantitative methods, other times through evocative metaphors – I wonder what all this intellectual labour amounts to in terms of the inevitability that faces the urban poor[1] of the world. Whatever the cause (development, infrastructural growth, gentrification, etc.), whatever the context (economic or non-economic), whether recorded quantitatively or not, *they must move*. Therefore, in the next section, I think back on my research on Ahmedabad city, India, conducted between 2006 and 2010, in an effort to direct intellectual energy into re-thinking how, as intellectuals, we can deliver a trenchant critique against the inevitability of displacement. I interviewed a hundred people living in the slums and slated for displacement along the banks of the Sabarmati river. Sixty-five of these were Muslim respondents and thirty-five were Hindus – more Muslims were willing to talk to us than Hindus because of the communally charged nature of Ahmedabad city following a Hindu-Muslim riot in 2002, in which the Hindus were the aggressors (Ahmed, 2002; Yagnik and Seth, 2002; Chatterjee, 2012). All interviewees were from separate households and belonged to the adult age group of twenty years and above. Some names of interviewees have been changed where anonymity was requested. The interviews were semi-structured and sometimes in the form of oral histories. I also interviewed twelve government officials – namely, the mayor, the deputy mayor, the district collector (chief bureaucratic executive), the commissioner and the deputy commissioner of the Ahmedabad Municipal Corporation (AMC), the chief executive authority, the ex-chief executive authority, and three other members of the Ahmedabad Urban Development Authority (AUDA), a district development officer, and the head of the slum clearance board. The interviews, oral histories, and ethnographic material were directed at understanding what Lefebvre (1970) calls the 'socio-logics' and 'ideo-logics' of urban renewal and the nature of disputes that emerged in response to those logics. 'Socio-logics' meaning the socio-spatial rationalities officially provided to implement urban renewal projects like the SRFD, and the 'ideo-logics' referring to the ideologies of entrepreneurial neoliberalism that actually underlie these socio-spatial rationalities. In the next section, using narratives from that research, I reconstruct the story of displacement in Ahmedabad city to provide a context for the critique that, I hope, will gradually emerge. My purpose here is not to construct a 'subaltern urbanism' or explicate how the megacity of the Global South represents a 'constitutive outside' (Roy, 2011), although these are important theoretical adventures. I accept here at the very outset that my training in Marx, Lefebvre, and other theories of the Global North travel with me to Sabarmati and inspire how I engage with the urban conditions there. However, I allow people's narrative to pollute, mould, shape, and re-inform my gaze from the North. The very heart of the people's narrative that informs my Sabarmati musings is that *the poor must move and they must resettle somewhere else*. The displaced self-conceptualize the material context of displacement and abstract it as an *inevitability*. It is in this self-conceptualization that I moor my conceptual musings to rethink how we do displacement theory by letting the displaced push the limits of Anglophone urban theory.

Displacement from the banks of the Sabarmati

The Sabarmati river cuts through the heart of Ahmedabad city in western India, dividing it into east and west. The east represents the older historical city, which was industrialized from the 1840s onward, becoming the basis for textile capitalism in India (Mahadevia, 2002; Chatterjee, 2016). Post-1980s structural adjustment and increasing global competition led to de-industrialization, mill-closures, increase in urban poverty, and increase in unemployment (Kundu and Mahadevia, 2002; Chatterjee, 2009). The working class and the growing informal proletariat are cleaved around ethno-religious lines, with the majority identifying as Hindus and the minority as Muslims concentrating around the now derelict mills. West Ahmedabad symbolizes the globalizing Ahmedabad, the 'world-class' city replete with malls, hotels, theatres, and theme parks inhabited by middle class and rich Hindus. With the post-1980s turn towards neoliberal urban entrepreneurialism, the AMC has paid special attention towards discursively and materially altering Ahmedabad's image from a declining mill town to an emerging market and a safe destination for global investments. Neoliberal entrepreneurialism manifests in the form of various greening, cleaning, road widening, and mall building projects, much of which involves reclaiming land from slum-dwellers and working-class neighbourhoods (Mahadevia, 2002).

An important flagship project of the AMC that showcases its entrepreneurial turn and its zealous adoption of the urban renewal goals is the Sabarmati River Front Development (SRFD) project. The need to excavate the Sabarmati river from its stagnant natural state into an unnatural spectacle of 'space, light, exhilaration' was fuelled by a desperate entrepreneurial urge to become 'world class' like Sydney, London, or New York – cities that are anchored by an urban imaginary of water.[2] The AMC launched its ambitious SRFD project at an estimated cost of USD 262 million, and constructions began in 2004. The project extends over 20.4 km along the Sabarmati, 9 km on one bank and 11.4 km on the other. The aim was to reclaim 162 ha of riverbed, fill it, and raise it to ground level. The land was then sold to private developers for building hotels, gardens, boulevards, promenades, water sports, heritage parks, skyscrapers, malls, and a straight road to the airport for high-speed travel (Chatterjee, 2009). An independent agency consisting of officeholders of the AMC has been registered as a company under the name Sabarmati River Front Development Corporation Limited (SRFDCL). SRFDCL was given the right to implement the project.

However, the project was soon mired in controversy because it entailed the removal of thousands of low-income families (both Hindus and Muslims) who inhabited the banks of the river. These are the urban working class; many had parents or grandparents that once worked in the textile mills, and many were migrants from surrounding rural areas displaced by agricultural poverty. In the context of contemporary de-industrialization, most of these families did informal work like hawking, vending, construction work, rickshaw pulling, domestic help, or chauffeuring (Chatterjee, 2009). The nature of their informal work required that the informal working class live in the heart of the city alongside the middle class and the rich who employed them. The banks of the Sabarmati, cutting through the heart of the city and constituting unused government land, served as an ideal place for squatting and slumming: they were strategically close to the centre but for a long time unclaimed by urban renewal because the ground was muddy and flood-prone during monsoons and cracked and parched during the dry season. Most settlers have been living on the banks for over forty years, and some moved in during the 1980s. Life along the Sabarmati river was a vibrant landscape of productivity, innovativeness, and resilience; at the same time, it is a landscape of poverty, marginality, and informality.

The AMC had employed Nilima Rawat, a consultant, to supervise the relief and resettlement efforts. Rawat claimed that about 8,000 families were displaced, but she also clarified that nongovernmental organizations (NGOs) put the figure at 30,000. The leader of the Sabarmati Nagrik Adhikar Manch (SNAM), the largest social movement spearheading resistance to the project, counterclaimed that there were 38,000 families along the bank of the river and that while the AMC claimed that only 8,000 were affected by the project, in reality, at least 20,000 families were moved. Nine bridges dissect the river and its banks into pockets of habitation, and these pockets have tended to become ethno-spatially distinct – with Muslims and Hindus in distinctly separate pockets. The Hindus were not a homogeneous mass; their habitation areas were subdivided into pockets of migrants from other states and those from Ahmedabad-Gujarat and pockets that were distinctly homogeneous in terms of caste. In terms of class position, they were all poor, engaged in informal work. The Muslim inhabitants and their living conditions were distinctly more desperate. Although, at a cursory glance, life along the river appeared to be random and chaotic, with people squeezing in to find space for survival, a closer look revealed the organized nature of the slum landscape, with the bridges spatially compartmentalizing religious and caste groups so that they lived in tightly knit communities alongside one another. Each group was keenly aware of the local micro-geographies – knowing exactly where one caste-based neighbourhood ended, where another immigrant community started, or where the Muslim *patti* (neighbourhood) began. The homes ranged from one-story brick-and-mortar rooms with terracotta roofs to flimsy huts made of bamboo, discarded wood pieces, and mud with plastic or tarpaulin sheets for roofs.

By 2010, some 1,300 families had been displaced, and the authorities claimed that people were moving because they want to move. The AMC and its associates claimed that large numbers of slum-dwellers got themselves listed for displacement because they wanted to have a better home, with a bathroom instead of the river and a kitchen instead of a cooking corner. The only challenge was to coordinate an appropriate resettlement strategy and organize transport since jobs were expected to germinate naturally in resettlement sites. After all, the poor are really innovative; as their skill-base is porous, they can effortlessly move from one kind of work to another. For MPK and SNAM, displacement is a continuous journey of the poor class – an empowering mobility, a habitual circulation engrained in the genetic code of the poor, a giving-up of space to create a better city (a better space), as long as new spaces are assured. The poor must flow through space. The *flowing* is assumed to be progressive for the city and the poor. On the other hand, for many inhabitants along the banks of the Sabarmati, like Kurshidbhai, displacement is injustice. Kurshidbhai and his group claimed that the AMC initially promised to build subsidized homes for the poor along the banks of the Sabarmati so that they need not be displaced. However, once the price of land along the banks started soaring, the AMC abandoned its original plan and decided to displace the poor to far-off mill lands, often outside the city limits. For many along the Sabarmati, displacement lead to a future of separation, otherness, and joblessness.

Another factor complicating displacement is the strained intercommunal relations between the working-class Hindus and Muslims since the 2002 riot days. Systemic Islamophobia perpetrated by the Bharatiya Janata Party (BJP), the ruling party in Ahmedabad at the time, led to the killing of thousands of Muslim minorities and the destruction of their homes, with many fleeing and being ghettoized in all-Muslim neighbourhoods outside the city (Ahmed, 2002; Yagnik and Seth, 2002; Chatterjee, 2012). In that post-riot strained intercommunal context, displaced Hindus and Muslims did not want to be resettled in close proximity to each other. Hindus felt that the Muslims did not deserve resettlement because they were 'troublemakers', and the Muslims feared an unfair resettlement process given the institutional Islamophobia. In March of 2010, some Muslim families slated to be displaced to the Vivekananda Mill complex in Rakhyal (in the extreme south-eastern periphery of the city) went to take a look at the new homes allotted to them by lottery. The Vivekananda Mill complex is one of the old textile mill complexes that were once the mainstay of Ahmedabad economy and that had been reclaimed by the government after they closed down in the late 1980s. These mill complexes, dotted around the city periphery as well as outside the city limits, have been converted into resettlement sites. When the Muslim families reached the resettlement sites, they were greeted with sticks and stones and angry Hindu mobs threatening to kill them if they ever came to claim these new homes (Chatterjee, 2014). In the nearby Ajit Mill area, which is predominantly Muslim, erstwhile mill land was converted into 700 resettlement homes, of which 300 were allotted and 65 more were slated to be allotted to Hindus. Muslims residing in Ajit Mill claimed that there was no intercommunity problem, and the Hindus who have moved to this mill complex were happy. I spoke to a Hindu family, and the head of the household complained bitterly that the resettlement site was far away from his business network. He also complained that the resettlement was conducted in a haphazard and insensitive manner – often ten or more people were crammed into two-room homes. Most importantly, he regretted having moved to a predominantly Muslim neighbourhood; he explained that the other Hindus who had been allotted homes in the Ajit Mill resettlement complex did not move in because they were too scared to live among Muslims; most moved in with relatives elsewhere and occasionally visited their allotted homes.

The resettlement of the Sabarmati inhabitant became a complex task of renegotiating the micro-geographies of ethnicity, fear, and intolerance. Spatial reasoning should have been deftly applied through statistically assumed neutrality that randomly allotted homes; however, this spatial reasoning should have also been deftly combined with a sensitive understanding of the ethno-geography, the fear topologies, and the thresholds of intolerance in order to peacefully reorder urban life. By 2014, the banks of the Sabarmati had been cleared off their informal proletariat and transformed into an awe-inspiring spectacle of terraces, parks, promenades, eateries, cafes, malls, jogging parks, and boat rides. The urban poor moved to make this possible. In the next section, I address this Sabarmati story in the context of MPK's statement that displacement is an inevitable reality of a poor person's life. Thus, I aim to critique displacement by taking my cues from the field. By using people's narratives front and centre, I aim to excavate concepts that were self-conceptualized by the displaced. These self-conceptualizations, I argue, allow us to rethink how we use displacement theory by letting the displaced push the limits of theory. Development-induced, gentrification-induced, revitalization-induced, quantitative, and qualitative assessments are useful academic/theoretical abstractions emanating from concrete realities of displacement. Nevertheless, I have not met anybody on the banks of the Sabarmati that described their own condition as 'development-induced' or 'gentrification-induced' or 'revitalization-induced' or described themselves as 'quantitatively' significant so as to demand resettlement. Therefore, I seek to abstract from the concreteness of the inevitability of displacement to produce counter 'socio-logics' and 'ideo-logics' so that 'the urban practice of the displaced, their way of life, their morphology' can confront elitist urban ideology (Lefebvre, 1970: 136–137). In this way, disputes would evolve into demands even when the revolutions themselves sometimes internalize the inevitability of displacement: 'the poor will have to move' (MPK, 2006 interviews).

Thinking back – they must move and resettle somewhere else...

Thinking back on the specific case of displacement and resettlement I have visited and written about in the past ten years (Chatterjee, 2009, 2012, 2014, 2016), it occurs to me that we have been looking at it from the wrong angle. Those that must move do not really care what caused displacement (development, gentrification, or urban renewal without gentrification), what was the nature of displacement ('physical' through bulldozing of homes or 'economic' through rise in rent), in what scale it occurred (urban, state-wide, provincial, or regional), whether they can be measured or not, or whether they moved a few miles or thousands of miles. What matters to them is that displacement embodies suffering and vulnerability in terms of loss of livelihoods, pangs of hunger, increase of poverty, and absence of basic amenities (like drinking water, schools, graveyards, etc.). Therefore, as humans who are also intellectuals, how can we best conceptualize this displacement so as to critique it comprehensively? Here, I argue for conceptualizations that move away from classifications based on causality, nature, measurement/methodology, and degree/extent. Instead, taking my cue from MPK, I focus on what is inevitable in the process of displacement. Based on my Sabarmati musings, I argue that the inevitable component of all displacements is that *the poor must move and they must resettle somewhere else*. It is here that I want to moor my conceptual musings, aiming to convey the human component of exploitation embedded in displacement. Such conceptualizations must emerge out of a dialectical relationship between the displaced and their abstraction of reality, not pigeonholed by the textbooks we carry to the field. Failing to do so would mean inter-neighbourhood comparisons to ensure whether gentrification-induced movement was significant compared to non-gentrification (Freeman and Braconi, 2004). Failing to do so would mean a discrete partitioning of reality so that displacement becomes 'secular replacement' (Hamnett, 2003), seen as a normal occupational transition as, for example, London becomes post-industrial. Failing to do so would also mean that no matter how poignant the tale of displacement is, as long as we are unable to produce an exact headcount, it will remain invisible within planning documents (agreeing with Slater, 2006). Failing to do so would also mean that I am embroiled in measuring the 'social cost' (Atkinson, 2004) of displacement rather than the exploitation it embodies.

Following Lefebvre (1970), I argue that ideological urbanism manifested in the SRFD project politically declares displacement as inevitable for the city's globalizing image. The soon-to-be displaced dispute this ideology and create urban confrontation and beginnings of 'concrete democracy', but this urban revolution is incomplete as the disputants internalize the urban ideology of the inevitability of displacement. What is politically/ideologically declared as inevitable becomes ultimately inevitable as the disputants, despite sizable resistance, self-conceptualize their displacement as inevitable. The following are three possible starting points for conceptualizing this inevitability.

Displacement as class exploitation

If the displaced, or soon-to-be displaced conceptualize their reality as *the poor must move and resettle elsewhere*, then even though we have not measured them, identified causality, and estimated the degree of their displacement, displacement has still happened. Whatever the cause, whatever the extent, whether we can measure it or not, the inevitable juxtaposition of displacement and class position *self-conceptualizes* the essence of exploitation embedded in displacement. I can take my cues from there and see how the inevitability of class exploitation is spatially played out in the banks of the Sabarmati and elsewhere. In the Sabarmati case, the soon-to-be displaced comprised people who are doing informal labour, self-employed, and unemployed. These were the urban poor, squatting on government land, but they were a productive force nonetheless – keeping the urban machinery alive through care giving, vending, transport, and custodial work. The bulldozers did not go to the middle-class and elite neighbourhoods of West Ahmedabad; they went to the banks of the Sabarmati – not because the inhabitants were unproductive to the urban accumulation process but because they were the poor class. Being poor means lacking capital (economic, political, and cultural) and, hence, being susceptible to exploitation manifested as inevitable forced displacement. If the displaced self-identify as a *poor class*, then that is the conceptual site that should matter intellectually in our schema for displacement. We must ask why certain classes experience the inevitability of displacement while others are resistant to it. In other words, how does class inflect with space so that the latter is rendered fleeting to some, making displacement an inevitability? My respondents lived along the banks of the Sabarmati in rows of shanty settlements as a class of informal poor until capital 'discovered' the 'badlands' along the banks of the river and rendered their right to space fleeting. As 'ideo-logics' of neoliberal urban renewal flow through the cities of the Global South, spaces of the poor class will be reclaimed for profit one after the other unless the poor – through their way of life, their morphology, and their conceptualization of displacement – convert disputes into demands and challenge that ideology through urban revolution. Nobody displaces the middle-class and rich living in condos in West Ahmedabad; displacement is not inevitable for the elite class. Displacement is always an ideology of class-space exploitation and, therefore, must be conceptualized in a class context. The nature of this class may be different in different settings: it could be in the form of urban informality, or rural peasants, or indigenous populations, or 'financification' (Lees, 2000: 392), where highly paid white-collar workers in the financial sectors of London and New York displace low-income groups. If class inevitably limits the right to say put, then displacement is *inevitable* only because *class exploitation* exists. An emancipatory urban process cannot address displacement piecemeal without addressing wage relations and inequality within capitalism. In other words, the right to challenge the inevitability of displacement is predicated on the urban poor's ability to challenge informality, unemployment, and semi-employment by demanding a just redistribution of the social product. Therefore, our conceptual critique must include at the forefront what the displaced on the banks of Sabarmati have laid out for us: a class analysis of the inevitability of displacement.

Displacement as race

Often, the poor that must move also encompass racial, religious, or ethnic minorities or immigrants. In the SRFD case, Muslims feared that there will be injustice embedded in the resettlement process, and when resettlement does happen, the new threads of social life will be hard to bind because of the ethno-religious biases embedded in urban planning. In fact, in my interviews, many Muslim households referred to this project as 'Muslim removal' because a large proportion of the inhabitants along the banks of the Sabarmati were poor Muslims. Given the post-riot sensitive relationship between the Hindus and the minority Muslim community, MPK claimed that in his negotiation for just resettlement, the Muslim people of Sabarmati have asked for separate resettlement sites for the Muslims.[3] In the context of gentrification, Lees (2000) argues that despite a burgeoning interest in urban marginality, very little work actually studies the intersection between race and gentrification. She argues that the intersection between class, gender, and gentrification is much more ubiquitously explored than empirical studies of race. Pulido (2000) uses examples of environmental racism to argue that urban research must not only explore instances of explicit intersection between racist policy and spatial transformation but also pay equal attention to the racism implicit in events like suburbanization and White flight from inner cities in the United States.

There is very little academic work that traces the ethno-religious or racial components of urban displacement. In Smith's (1986) book on gentrification, race is the underlying solvent that dissolves the Latino population in the Lower East Side. But a Google Scholar search on 'displacement' and 'race' yields almost nothing. The overt and covert imbrication of race and ethno-religiosity with the inevitability of *they must move* must be conceptually excavated. Who gets to stay, who has the right to stay put, is as much inscribed on the skin or ascribed to religious identity as it is inflected by class position. Therefore, if we begin with the displaced or soon-to-be displaced and understand displacement as their narrative, we may find that the inevitable attribution of 'must move' is not only related to poverty (class position) but often related to ethno-religious/racial positions: 'because we are Muslims, the government does not care about us, and they want to remove us from predominantly Hindu neighbourhoods in the centre of the city'.[4] It is possible that race-displacement may not be as overtly visible or as empirically fortifiable as class displacement. Planning policy may clearly identify indicators of blight and lay down policy action in the context of eminent domain related to blight-like situations, which may in turn enable us to conceptually connect blight with poverty, low-income, and hence, displacement. It may be harder, but not impossible (Jackson, 1987; Baeten, 2001; Corburn, 2009), to distinguish racism in planning policy documents. Racism is contextualized differently depending on the 'socio-logics' and the 'ideo-logics' of 'urban ideology' (Lefebvre, 1970: 136–137). Therefore, in the Sabarmati case, it must be understood through the lens of ethnoreligiosity and Islamophobia, which do not manifest in skin colour but still determine the topography of displacement as inevitability – yet another reason why urban theory from the North cannot be un-problematically superimposed on the South. In the Sabarmati context, the AMC emphatically denied any collusion with right-wing Islamophobic politics. Nevertheless, ethno-religiosity was ever-present in the way the soon-to-be displaced imagined their resettlement, in the way in which the AMC was rethinking the re-settlement efforts, and in how MPK was re-envisioning 'displacement with dignity'.[5] Therefore, I argue that even in the absence of statistical significance, or even in the absence of racist policy documents, if ethno-religiosity, race, caste, and indigeneity are present in the inevitable tropes of why *they must move*, it is our responsibility as researchers to tease out the implications of race, religion, ethnicity, caste, indigeneity in shaping displacement.

Displacement as resettlement

In rethinking the Sabarmati displacement story, it also occurred to me that displacement is not the beginning and end of exploitation. The displaced must re-negotiate new life in the re-settlement site and claim the randomly allotted home many miles away from where they have lived forever. Sometimes the resettlement site is outside the city limits, with scarcity of basic amenities such as public transport routes, schools, drinking water, and graveyards. Sometimes the resettlement site is an overcrowded room among hostile neighbours who threaten to kill you if you dare to re-settle. If displacement exaggerates class and race exploitation, then resettlement gives it a new lease of life. Exploitation does not end with resettlement; instead, new topographies of marginality are etched out after *they have moved and resettled elsewhere*. The AMC in their first master plan had no intention of resettling anybody since the residents of Sabarmati were illegal squatters. However, the history of displacement around the World Bank funded Sardar Sarovar dam that had taken place in the 1990s in the state of Gujarat, where Ahmedabad city is located, and the social movement of the poor surrounding that dam-displacement were at the back of everyone's mind. The Narmada Bachao Andolan (NBA) that spearheaded the social movement against the dam project consisted of indigenous people, environmentalists, farmers, and human rights activists agitating against displacement through civil disobedience, demonstrations, and hunger strikes – starting an era of critique of modernity and development (Roy, 1999).[6] Nobody wanted a repeat of history; hence, the AMC introduced re-settlement planning. From the AMC's perspective, they were helping the displaced through adequate resettlement packages. For the inhabitants of the Sabarmati, resettlement opened new introspections of inevitability: that the poor must not only move, but they also do not get to decide what happens the day after displacement, and they must resettle elsewhere.[7] Elsewhere means exclusion. Marcuse (2013) argues it means physical and economic expulsion. Elsewhere often means nowhere – a complete loss of control over the urban process. The inevitability of class/race exploitation is embedded not only in displacing but also in the resettlement process. The resettlement process is supposed to attribute justice to the displaced and, therefore, characterize a due process that makes displacement humane. On the other hand, the Sabarmati story indicates that resettlement is not a happy place of emplacement, because the exploitation of displacement continues to impinge on resettlement sites. Atkinson (2000: 163) characterizes the problem aptly: 'Displacement removes social problems and rearranges rather than ameliorates the causes of poverty, environmental decay and the loss of neighbourhood vitality – problems are moved rather than solved'. The people inhabiting the banks of the river Sabarmati represent displaced labour from rural areas and displaced labour from formal sectors of the economy (now-closed textile mills). The reason for squatting along the banks of the Sabarmati is the super-exploitative wage that is sporadically earned in the informal sector. With these below-subsistence wages, the people of Sabarmati can either eat or pay rent, not do both; they choose to eat, so squatting and slumming become the only alternative for existence. Tackling informality through employment planning and wealth re-distribution could mitigate urban inequality. However, the ideo-logics of urban renewal pose only superficial solution through spectacular projects and associated resettlement of the poor. The resettled poor continue to be poor in the peripheries of Ahmedabad, facing even bigger impediments to their livelihood. Displacement is the material representation of the inherent class and racial contradictions of capitalism, and resettlement is posed as a superficial solution to 'manage' that contradiction instead of fundamentally altering wage, class, racial relations and hence altering the inevitability of marginality: *they must move and resettle somewhere else.*

Conclusion

In this article, I have attempted to think back on my research on displacement surrounding the SRFD project in Ahmedabad city conducted between

2006 and 2010. My effort has been to understand how the field, the lives, and the narratives of displacement and resettlement inform my intellectual introspections. In the gerbil-wheel of academic publishing, the field, and the interviewees, the 'subject' and the statistics are often a means to an end; that is, they must tell us something that fits within the existing body of literature, and preferably, they must advance the existing body of literature by either proving or disproving something. In that context, case studies from the Global South are valuable as they are supposed to confer a touch of worldliness to the fixed attributes of set concepts that emerged in the Global North. Therefore, I revisited my Sabarmati field notes and juxtaposed them against a wide array of existing theories on displacement – many from the Global North, although not exclusively so. I was struck by the number of groups, debates, and classifications within displacement theory. Gentrification was the dominant theme in displacement literature in the Global North, and infrastructural development the dominant theme for displacement in the Global South. Then there were 'smaller' themes, like urban renewal, revitalization, neoliberalism, colonization, and capitalist-urbanization as important displacement inducing factors. These were sub-stratified into those that did quantitative analysis and those that did not. Another prominent intellectual undercurrent (within the Global North) was the claim that because of the want of empirical data, major propositions about displacement must be tempered with caution. I realized that as a person born in the Global South but academically trained in the Global North, I have the advantage of a compound gaze – I carried with me the 'baggage' of Anglophone urban theory, but I was now capable of juxtaposing urban realities from the Global South against the theories of the Global North (many of which are quite inspiring and useful) to re-conceptualize urban reality.

Thus, I wondered if it was possible to use people's narratives as front and centre, take my cues from the field, and excavate concepts self-conceptualized by the displaced themselves to represent the inevitability of displacement. Inspired by Lefebvre, but un-inspired by the causal conceptual categories that I had gleaned from Anglophone and non-Anglophone urban theory (Development-induced, gentrification-induced, revitalization-induced), I decided to shed the latter and instead use the conceptual abstractions that emerged from the displaced person's self-conceptualization of the material contexts of displacement. Accordingly, I thought, could I use the field and start to rethink and push the limits of how we do displacement theory by letting the displaced encapsulate us completely? I used displacement narratives to conceptualize the *inevitability* of displacement that is palpable in all the Sabarmati narratives, in which the poor contends *that the poor must move, and they must resettle elsewhere.* I treat this not as a statement from an interviewee in the field but, rather, as a set of concepts that provided entry points to help us understand displacement fully.

Tracing this set of concepts as the underlying foundation for this article, I arrive at three lenses that the displaced on the banks of the Sabarmati provide: displacement as class, displacement as race, and displacement as resettlement. The displaced explicate how urban exploitation is almost always class-based, almost always an attack on labour, and almost always an erasure of the poor and their geographies. The displaced also explicate how displacement is shaped through minority-othering and Islamophobia. In fact, religiosity becomes a major guiding force for structuring resistance to displacement and planning resettlement. Lastly, the displacement narratives indicate how exploitation is extended the day after displacement through resettlement. Resettlement as a process is conceptualized by the displaced as government-neglect, marginalization, and spatial-outcasting. The displaced along the Sabarmati provide substantive entry points to a theory on displacement that begins with the displaced. If Lefebvre (1970: 137) is correct that disputes reveal the ideological characteristics of groups and classes and if elitist urban renewal projects like the SRFD represents the elitist ideological-morphology, then the fact that the largest resistance movement there does not confront the very heart of urban ideology (i.e., displacement) must have something to tell us. It says that the urban poor's 'way of life' (Lefebvre, 1970: 137), their slum morphology-ideology, assimilates ideological urbanism without disputing its *ideological* basis because it accepts displacement as *inevitable*. What is politically declared as inevitable (ideology) becomes inevitable (material reality) because of the juxtaposition of class, Islamophobia, and the spatial politics of resettlement. The gist of displacement theory from the Sabarmati narratives is that displacement is inevitable for the poor and the ethnic-minority, and it acquires resilience through resettlement. This self-conceptualization informs my North–South compound gaze and hopefully disrupts, bleeds into, remoulds, and pollutes existing cause-effect categories. The purpose here is conceptual destabilization so that emerging critique of displacement attacks class, racial, and spatial contradictions of the urban process.

Notes

1 I prefer the terms *urban poor* and *urban poverty*, inspired by Mike Davis's (2004: 11–13) characterization of the informal working class in the cities of the Global South as the 'new urban poor' or the 'outcaste proletariat', because it emphasizes a distinct characteristic of urban poverty in the Global South – that it is somewhat a product of rural to urban migration but largely a result of massive shrinkage in formal jobs in the neoliberal period (1980s) due to structural adjustment and de-industrialization (Chatterjee, 2009). This urban poor, Davis argues, has no stake in neoliberal accumulation and is therefore likely to be ripe with revolutionary potential.

2 Based on interviews of AMC officials, 2006.

3 2010 interviews.

4 Muslim interviewee in Khanpur, Ahmedabad, slated to be displaced in 2010.

5 Interview with Mohammad Pathan Aliyar Khan (MPK), leader of Sabarmati Nagrik Adhikar Maanch (SNAM) in 2006 and 2010 where he claimed that an additional criterion must be added to the resistance charter of demands that will ensure a dignified resettlement process where Hindus and Muslims are allotted homes in separate resettlement zones.
6 Interview with Jagdish Patel, Sabarmati Riverfront Development Corporation Limited (SRFDCL), 2006.
7 Interview with Kursheed bhai, Khanpur, Ahmedabad city, 2010.

References

Ahmed, R. (2002). Gujarat violence: Meaning and implications. *Economic and Political Weekly*, 18 May. Available at: http:// www.epw.org.in/showarticles.php [Accessed 5 April 2005].

Atkinson, R. (2000). Measuring gentrification and displacement in Greater London. *Urban Studies*, 37(1), pp. 149–165.

Atkinson, R. (2004). The evidence on the impact of gentrification: New lessons for the urban renaissance? *European Journal of Housing Policy*, 4(1), pp. 107–131.

Baeten, G., (2001). Clichés of urban doom: The dystopian politics of metaphors for the unequal city – A view from Brussels. *International Journal of Urban and Regional Research*, 25(1), pp. 55–69.

Bunnel, T., Barter, P. A., and Morshidi, S. (2002). City profile Kuala Lumpur metropolitan area: A globalizing city region. *Cities*, 19(5), pp. 357–370.

Carl LeVan, A. and Olubowale, J. (2014). 'I am here until development comes': Displacement, demolitions, and property rights in urbanizing Nigeria. *African Affairs*, 113(452), pp. 387–408.

Cernea, M. M. (2000). Risks, safeguards, and re construction: A model for population displacement and resettlement. In: *Risks and reconstruction: Experiences of resettlers and refugees*. The World Bank.

Chatterjee, I. (2009). Social conflict and the neoliberal city: A case of Hindu-Muslim violence in India. *Transactions of the Institute of British Geographers*, 34(2), pp. 143–160.

Chatterjee, I. (2012). How are they othered? Globalization, identity and violence in an Indian city. *The Geographical Journal*, 178(2), pp. 134–146.

Chatterjee, I. (2014). *Displacement, revolution, and the new urban condition*. Delhi, London, New York: Sage.

Chatterjee, I. (2016). Beyond the factory: Struggling with class and class struggle in the post-industrial context. *Capital and Class*, 40(2), pp. 263–281.

Clark, D. (2009). Power to the people: Moving towards a rights-respecting resettlement framework. In: *Development and dispossession: The crisis of forced displacement and resettlement*. Santa Fe, NM: School of Advanced Research Press, pp. 181–199.

Corburn, J. (2009). *Toward the healthy city: People, places, and the politics of urban planning*. Cambridge, MA: MIT Press.

Davis, M. (2004). The urbanization of empire: Megacities and the laws of chaos. *Social Text*, 22(4), pp. 9–15.

De Wet, C. (ed.). (2006). *Development-induced displacement: Problems, policies and people*. New York: Berghahn Books.

Doshi, S. (2015). Rethinking gentrification in India: Displacement, dispossession and the spectre of development. In: *Global gentrifications: Uneven development and displacement*. Bristol, UK: Policy Press, pp. 101–120.

Escobar, A. (2003). Displacement, development, and modernity in the Colombian Pacific. *International Social Science Journal*, 55(175), pp. 157–167.

Fisher, W. (1995). *Towards sustainable development: Struggling over India's Narmada River*. New York: M.E. Sharpe Inc.

Freeman, L. and Braconi, F. (2004). Gentrification and displacement: New York City in the 1990s. *Journal of the American Planning Association*, 70(1), pp. 39–52.

Ghertner, D. A. (2014). India's urban revolution: Geographies of displacement beyond gentrification. *Environment and Planning A*, 46(7), pp. 1554–1571.

Haldar, T. and Abraham, V. (2015). Development, displacement and labour market marginalisation: The case of Jharkhand tribal population. *Social Change*, 45(1), pp. 45–66.

Hamnett, C. (2003). *Unequal city: London in the global arena*. London, New York: Routledge.

Jackson, K. T. (1987). *Crabgrass frontier: The suburbanization of the United States*. New York, Oxford: Oxford University Press.

Koenig, D. (2006). Enhancing local development in development-induced displacement and resettlement projects. In: de Wet, C. (ed.) *Development-induced displacement: Problems, policies and people*. New York: Berghahn Books, pp. 105–140.

Kundu, A. and Mahadevia, D. (2002). *Poverty and vulnerability in globalizing metropolis: Ahmedabad*. New Delhi: Manak Publications.

Lees, L. (2000). A reappraisal of gentrification: Towards a 'geography of gentrification'. *Progress in Human Geography*, 24(3), pp. 389–408.

Lees, L., Shin, H. B., and Morales, E. L. (eds.) (2015). *Global gentrifications: Uneven development and displacement*. Bristol: Policy Press.

Lefebvre, H. (1970). *La révolution urbaine* (Vol. 216). Paris: Gallimard.

Marcuse, P. (1986). Abandonment, gentrification, and displacement: The linkages in New York City. In: Smith, N. and Williams, P. (eds.) *Gentrification and the city*. London: Unwin Hyman, pp. 153–177.

Marcuse, P. (2013). Abandonment, gentrification, and displacement: The linkages in New York City. In: Smith, N. and Williams, P. (eds.) *Gentrification of the city*. London, New York: Routledge, pp. 169–193.

Mahadevia, D. (2002). Communal space over life space. *Economic and Political Weekly*, 30 November. Available at http://www.epw.org.in/showarticles.php [Accessed 20 April 2005].

Marinelli, M. (2009). Negotiating Beijing's identity at the turn of the twentieth century. In: Butcher, M. and Velayutham S. (eds.) *Dissent and cultural resistance in Asia's cities*. London and New York: Routledge, pp. 33–52.

Nayak, A. K. (2013). Development, displacement and justice in India: Study of Hirakud Dam. *Social Change*, 43(3), pp. 397–419.

Oliver-Smith, A. (ed.) (2009). *Development and dispossession: The crisis of forced displacement and resettlement*. Santa Fe, NM: School for Advanced Research Press.

Parasuraman, S. (2016). *The development dilemma: Displacement in India*. London: Macmillan Press.

Patel, S., Sliuzas, R., and Mathur, N. (2015). The risk of impoverishment in urban development-induced displacement and resettlement in Ahmedabad. *Environment and Urbanization*, 27(1), pp. 231–256.

Pulido, L. (2000). Rethinking environmental racism: White privilege and urban development in Southern California. *Annals of the Association of American Geographers*, 90(1), pp. 12–40.

Roy, A. (1999). The greater common good. Available at: http://www.narmada.org/gcg/gcg.html [Accessed September 2018].

Roy, A. (2011). Slumdog cities: Rethinking subaltern urbanism. *International Journal of Urban and Regional Research*, 35(2), pp. 223–238.

Slater, T. (2006). The eviction of critical perspectives from gentrification research. *International Journal of Urban and Regional Research*, 30(4), pp. 737–757.

Smith, N. (1987). Gentrification and rent gap. *Annals of the Association of American Geographers*, 77(3), pp. 462–478.

Smith, N. (1996). *The new urban frontier: Gentrification and the revanchist city*. New York: Routledge.

van Cleef, A. (2016). Hydropower development and involuntary displacement: Toward a global solution. *The Indiana Journal of Global Legal Studies*, 23, p. 349.

Yagnik, A. and Seth, S. (2002). Wither Gujarat? Violence and after. *Economic and Political Weekly*, March. Available at http:// www.epw.org.in/showArticles [Accessed 3 April 2005].

Yeoh, B. and Chang, T. C. (2001). Globalizing Singapore: Debating transnational flows in the city. *Urban Studies*, 38(7), pp. 1025–1044.

3 A landscape of post-gentrification?

A renovation case in Sweden

Emil Pull

Introduction

In an article reflecting on the definition and scope of gentrification, Eric Clark (2005) recounts a meeting with pioneering gentrification scholar Neil Smith. Asked to be shown the battlefields of gentrification in the city of Malmö, Clark hesitated and found it difficult to explain that gentrification was indeed prevalent in Sweden, and had been for a long time, but the country had no proper battlefields. Over a decade later, the battlefields are here – or, at least, a hot glowing ember. Neighbourhood unrest is gaining attention in media, politics, and academia. Anti-displacement grassroots networks are growing, and articles about displacement and displacement trauma are prevalent in the news. This development can be predominantly attributed to a sharp rise in overhaul renovations of the large housing stock from the 1950s to 1970s and the accompanying rent increases for tenants. Nonetheless, is this really a new gentrification frontier we are witnessing in Sweden or a paradigm shift to a post-gentrification housing regime? This chapter will discuss these questions using the example of Kvarngärdet and Gränby, two Swedish neighbourhoods in Uppsala municipality, the fourth largest city in Sweden. The findings are built on the case work of the authors' dissertation thesis, in which forty interviews with tenants were carried out over a period of five years (2014–2019).

Kvarngärdet and Gränby

Kvarngärdet and Gränby are two downtown-adjacent neighbourhoods in the municipality of Uppsala, Sweden's fourth largest city. Kvärngärdet, in particular, is largely within walking distance of the central station and the commercial district of Uppsala. Gränby is located a bit further out and, though smaller in population, is larger in size and contains a lot of green areas towards the fields surrounding Uppsala (see Figure 3.1). Both neighbourhoods have had large parts of their housing stock renovated over the past decade and, particularly in Gränby, a lot of new multifamily homes have been (and are being) constructed in the last few years.

Figure 3.1 Map of Uppsala.
Image processed from Google Maps.

In a first step, early in 2001, the municipality-owned housing company Uppsala Hem announced the need to upgrade their housing stock in Kvarngärdet. These initial renovation plans were met with acclaim by the respondents, who were well aware of the acute need for renovation of their largely neglected buildings. My respondent Theresa told me,

> These houses were built during the 1960s to solve the housing shortage in Uppsala. They were built with lower standards and were only supposed to stand for twenty years or so and then be demolished. But after a while, they noticed they could stand for longer [...], but as an effect, these houses have never been taken care of. No changes of plumbing, electricity, or windows. Nothing! Just the other day, there was a smell of something burnt coming from my wall. It is the electricity.
>
> (Theresa, interviewed in January 2015)

However, nothing happened with the plans until a few years later, when the municipal housing company sold parts of its stock to a private landlord. Uppsala Hem then went forward with the renovation plans of their remaining stock, and the new landlord informed the tenants that their houses were scheduled to be renovated as well. However, there was a caveat: both landlords announced that the renovations would be financed through rent increases. Numbers as high as hundred per cent were announced. Since then, my respondents in both neighbourhoods have been fighting not only to be able to stay but also for the maintenance and renovations of their homes. Without renovation, as Karin (another respondent) puts it, their homes 'feel like a deathtrap'. Since 2015, I have returned over a dozen times to talk to my respondents. Today, they are tenants who have all suffered rent increases between eighteen and sixty-five per cent, and they are all tenants under various forms of 'displacement pressures' (Marcuse, 1985). These pressures are making their lives nearly unbearable with hardships and grief and also angry and with a lust for agency. The built-up pressures and the feeling of being neglected as valued members of society (Polanska and Richard, 2019) are of course spilling into the streets of the neighbourhood, and incidents of small riots have occurred. As one respondent, Karin (a pensioner aged seventy-three), puts it, 'Under the circumstances, I understand the youth torching cars here. Had I only been younger, I would have joined them myself!'.

Gentrification in a Swedish context

Sweden has a long history of gentrification, even if the various processes have not always been denoted as gentrification. In fact, gentrification as a concept was introduced in research and student literature relatively late in Sweden. This is possibly due to the fact that urban development processes played out differently than in the United States and the United Kingdom, for instance. There is a landmark study from 1987 by Eric Clark, where he investigates gentrification in the Swedish city of Malmö over the years 1860–1985. This was a pioneering work, both empirically and geographically, but it was not until the second decade of this century that scholars like the aforementioned Clark along with Hedin (2010) and Thörn and Holgersson (2014) translated gentrification into Swedish and popularized the concept – making it a household term among domestic researchers, students, practitioners, and laypersons. (It is still a relatively uncommon term in media and probably for the general public.) However, there are ample examples in Swedish post-war history where a gentrification lexicon can be used to illuminate various housing processes. The largescale and geographically diverse inner-city demolitions and modernization in the 1950s and early 1970s (Johansson, 1997) and the large-scale suburban renovations of the rental stock in the 1980s (Öresjö, 1993) are two such examples, where gentrification very well could, but rarely is, used to denote the processes of change. In 2012, Hedin et al. published a much referred-to study that found clear evidence of gentrification and super-gentrification in Sweden's three largest cities during two boom periods (1986–1991, 1996–2001) and a bust period (1991–1996). Much literature has focused on the neoliberal turn in housing politics and the consequential gentrification and socio-economic polarization of Swedish cities since then (e.g., Andersson and Turner, 2014; Sandberg, 2014; Baeten and Listerborn, 2015; Thörn and Holgersson, 2016). Nevertheless, in many ways Sweden is, or at least was, unique, having a tradition of social democratic housing politics and social engineering that often led to gentrifying processes without the classical disinvestment and neighbourhood decline cycles that much of the Anglo-American literature focuses on (Clark, 2005). Likewise, gentrification has been as often suburban as central. A long tradition of a large, egalitarian, high-quality, and universal public rental sector that deploys use-value baselines for rent-setting (where geographical location has not been an acceptable factor influencing rents for tenants, for instance) has kept rents generally affordable both in and out of the city centres (Hajighasemi, 2004). Strong legal tenant protections have made evictions hard and gentrification a relatively slow and mundane process. This has staved off much of the contestation that is often linked to gentrification in other countries and kept Neil Smith's battlefields at bay.

Much of this has been in the process of changing. The share of public housing is on the decline; it is being sold out to private actors or transformed into fully marketized condominiums (Salonen, 2015). Further, where they are still present, municipal housing companies are becoming increasingly inaccessible to poor dwellers as municipal companies increase access demands (e.g., minimum income levels and recommendations from former landlords) and produce housing with higher rents (compared to the disposable income of the urban poor) than ever before (ibid). Consequently, as Grander (2017) argues, the public housing of Sweden can perhaps no longer be called universal. It no longer caters for the segment of the population with the lowest incomes. In its place, there has been a radical growth in 'social housing contracts'[1] (nationwide, the amounts of social contracts have doubled in the last decade), to the extent that the municipalities' social services have become major players on the housing scene in many cities. Social housing contracts still only amount to a mere two per cent of the total housing market, but for a country without a social housing sector and where the universal public rent sector is supposed to be just that, universal, its rapid growth is worrying.

Renovations

The mechanisms turning renovations of tenant buildings into a violent and disruptive process leading to large-scale displacements have been thoroughly discussed elsewhere (Westin, 2011; Molina and Westin, 2012; Baeten et al., 2017). In short, the exceedingly sharp rent increases are made possible through what can most aptly be described as a legal loophole. From a legal standpoint, structural renovations (plumbing, heating, insulation, etc.) under the Swedish use-value system cannot be grounds for rent increases, but recent structural renovations of the housing stock from the 1960s and 1970s have forced so-called luxury renovations of bathrooms, floors, etcetera upon tenants. These sorts of renovations are permitted if they are deemed to be in the interest not of the current tenant but of an abstract 'general tenant'. When individual tenants contest the renovations in the rent tribunal court, they lose nine out of ten cases in favour of this 'general tenant' and the 'common good'. This loophole has been previously used by the state to upgrade what was considered socially stigmatized suburbs outside of Stockholm (Öresjö, 1997), with displacement (or social mixing) as an outspoken goal. Nevertheless, in spite of the otherwise strong legal tenant protections in Sweden, never before has this magic trick of closing rent gaps reached this scale. Over a million dwellings were built during the building boom, and hundreds of thousands of them have recently undergone renovation, are currently undergoing renovation, or are in need of structural renovations (Baeten et al., 2017). The vast majority of these renovations are paid for by tenants through the curious practice of 'luxury renovations' (ibid). In tandem, Sweden is suffering a near-nationwide lack of affordable housing, with 240 municipalities (eighty-three per cent) reporting housing shortages in 2019 according to the National Board of Housing. However, the housing shortage is unevenly spread between tenure forms (with the largest scarcity being in the affordable housing sector) and place (with city central neighbourhoods in the larger municipalities being affected the most) despite a general construction boom over the last decade. This obviously has an effect on displacement patterns. According to the National Board of Housing, Building and Planning (Boverket, 2014), displaced tenants often end up economically and socially worse-off. The lack of affordable housing also hinders people from moving to a cheaper location in the first place, as is the case for some respondents in Gränby and Kvarngärdet.

Gentrification or not? Working through methodology

In the following, I attempt to make an ad hoc test of gentrification for the development in Gränby and Kvargärdet using common methodological approaches in identifying gentrifiable, gentrifying, and gentrified neighbourhoods in the literature. The reason is twofold. First, it is a way to describe what is happening in the two neighbourhoods in question and to show how far off the happy side of sushi and coffee shop gentrification these renovations are. Second, I want to check whether the lexicon of gentrification is at all suitable for these processes in contemporary Sweden or what we are seeing is something new.

The methodologies measuring gentrification are many, and one divide is based on whether one ascribes to consumption- or production-driven gentrification. While going into any depth about this theoretical rift is outside the scope of this chapter, suffice it to say (and vital for the conclusion of this chapter) that the common denominator in these theories is *gentrifiers*, irrespective of whether they are driving the process via preferences and consumption patterns or merely absorbing the increased land rents and the consumeristic landscape that closing the rent gap[2] produces in the aftermath. In gentrification theory, a gentry is always implied, whether they are the drivers of the process or merely its beneficiaries.

Identifying gentrification is commonly done via longitudinal census data. In a two-step process, this method first identifies the gentrifiable areas and then identifies the gentrifying or gentrified tracts. Typically, the literature labels any area gentrifiable when the average or median incomes are below some percentage of the metropolitan average (Bostic and Martin, 2003; Wyly et al., 2010) or below some absolute threshold (McKinnish et al., 2010). In a Swedish context, the discrepancy between the average household income in Kvarngärdet and Gränby compared to Uppsala municipality as a whole was fairly marked before the renovation process started (see Figures 4.2 and 4.3). These neighbourhoods would certainly be labelled gentrifiable according to this method. In a more methodologically sophisticated approach, Freeman (2005) labels an area gentrifiable if the average incomes are below the metropolitan average and if the area has suffered from disinvestment for the past twenty years. Disinvestment here means lower-than-average newly constructed housing. This method of identifying gentrifiable areas has the advantage of corresponding nicely with early conceptualizations of gentrification as cycles of investment and disinvestment (Smith, 1979; Marcuse, 1985). The question is whether disinvestment is so easily captured by simply counting newly produced buildings. Kvarngärdet and Gränby certainly lacked investments in newly produced buildings and, consequently, match the criteria, but disinvestment in the neighbourhoods was perhaps even more notable, particularly for the inhabitants, in the lack of building maintenance over the decade preceding the renovations (as recounted earlier in the chapter). In any case, disinvestment in terms of an absence of newly produced housing was present in both areas prior to the renovations. Thus, both areas can be labelled as 'gentrifiable' using both methodologies.

Other researchers have produced multi-variable analyses, notably, Hammel and Wyly's (1996) methodologically rich paper outlining different census variables that discriminate between gentrifying and non-gentrifying tracts. Another is Bostic and Martin's (2003) study, where they develop Hammel and Wyly's approach and list changes pertaining to gentrification using the following nine factors: share of residents with college degrees, growth in average family income, homeownership rate, share of population aged thirty to forty-four, poverty rate, percentage of White non-family households, share of Black population, share of managerial and administrative workers, and percentage of residents with at least some college education. However, while a strict analysis based on a multi-factorial census analysis would be interesting, fieldwork in the neighbourhoods in question suggests that 'time period' or 'end of the time period' variables do not adequately capture the situation in Kvarngärdet and Gränby since the areas are still in flux. Nevertheless, the methodological contributions of Hammel and Wyly and of Bostic and Martin do identify certain data sets that have strong correlations with gentrification.

In terms of transition from a gentrifiable to gentrifying or gentrified tract, Freeman (2005), in contrast to other studies, ignores income levels in the gentrified area in favour of college education, arguing that gentrification (in adherence to the origin of the concept) is more about cultural class transition than income levels. To him, gentrification has occurred if housing prices have increased and the percentage of residents with a college degree has increased above the metropolitan average. Thus, income is a factor only in identifying gentrifiable areas, not actual gentrification.

According to most methodologies of measuring gentrification, Kvarngärdet and Gränby can certainly be categorized as 'gentrifiable' neighbourhoods. But are they undergoing gentrification? An analysis based on income levels before, during, and after the renovations gives, at best, a very weak correlation between the renovations and rising income levels. Both areas started well below the municipal average, both had stagnating or even declining income levels during the actual renovations, and only Kvarngärdet has been trending (although weakly) upwards towards the municipal average in the years after the renovations (see Figures 3.2 and 3.3).

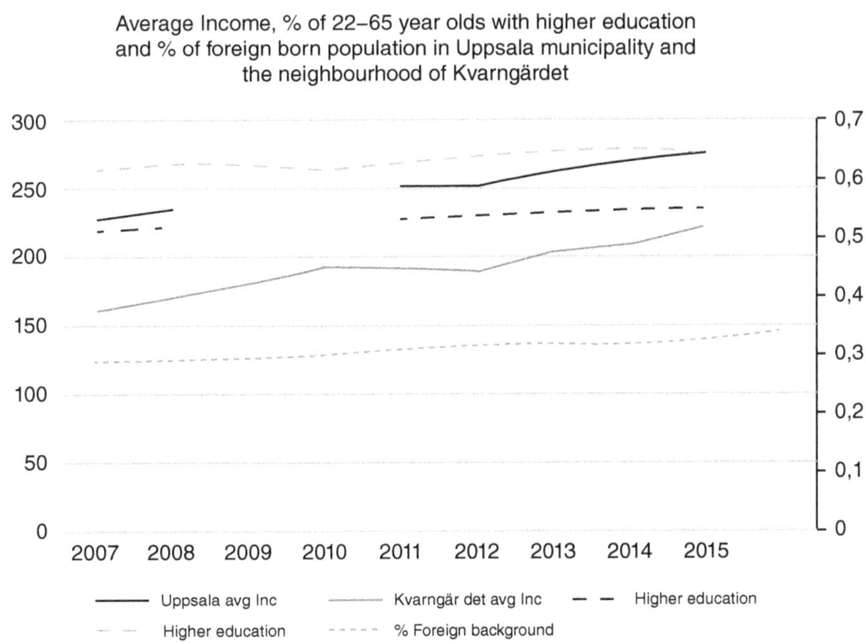

Figure 3.2 Gentrification indicators Kvarngärdet.
Source: Data from the Department of Urban Management, Uppsala municipality.

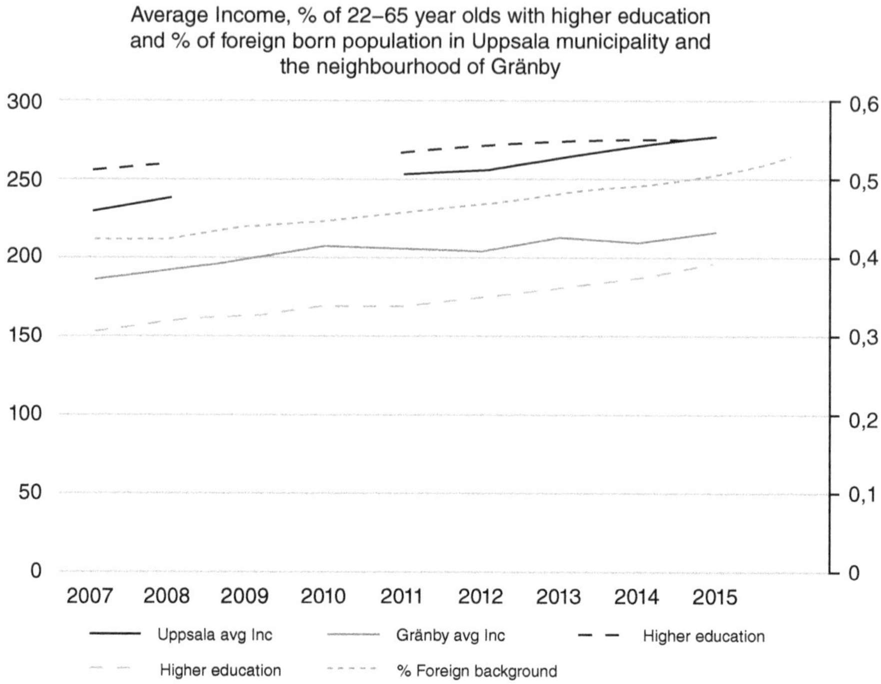

Figure 3.3 Gentrification indicators Gränby.
Source: Data from the Department of Urban Management, Uppsala municipality.

On the other hand, the census statistics correlate well with responses from tenants, who note that the affluent people moving into Kvarngärdet and Gränby are merely doing so on a short-term basis, on their way to someplace else in Uppsala. Asked directly if the neighbourhood is becoming a middle-class neighbourhood, one respondent answered,

> What!? God no! On the contrary. Any middle-class family that moves here moves out just as quick. There is garbage everywhere. No one cares about the neighbourhood. I would move if I could, but it's expensive everywhere. Most new people that move in are poor immigrants or on social housing contracts. And the poor immigrants aren't even getting information in a language that they would understand. Their rents are probably paid for by the municipality.
>
> (Birgitta, interviewed in March 2019)

If we consider education levels, another variable of Freeman's (2005) method, only Gränby would have been considered a gentrifiable tract prior to the renovations. Gränby has also seen an increase in education levels between 2007 and 2014 (see Figure 3.2), but the changes are very modest; in 2015, the levels were still well below the municipal average. According to the respondents, the small increase could be a result of students moving into the empty flats. But respondents both in Gränby and Kvarngärdet indicate that the neighbourhoods are stepping stones to more attractive living and that students only stay for a brief period. In contrast to Gränby, Kvarngärdet started out with above-average education levels and would therefore not be classified as a gentrifiable tract according to Freeman (2005). The education levels have also dropped in Kvarngärdet since the start of the renovations (see Figure 3.2). Thus, it is hard to see any gentrification in either of the neighbourhoods based on the shifts in education levels.

Homeownership rate is another factor used to gauge gentrification (Bostic and Martin, 2003). The changes in forms of tenure are relatively undramatic in Gränby and Kvarngärdet. In both tracts, the number of ownership/condominium dwellings have remained well below the municipal average (see Figure 3.4). The increasing share of ownership/condominium dwellings in Kvarngärdet can be chiefly explained by the newly built condominiums on virgin soil in the outskirts of the neighbourhood. This is a caveat, and it could possibly indicate long-term gentrification or new-built gentrification, but at least thus far, it has had little impact on the overall class-composition of the neighbourhoods.

A category seldom used in gentrification analysis in a Swedish context is ethnical or racial composition. One reason is that despite a decidedly racist dimension to housing provision in Sweden (Molina, 1997), the demarcation between White and Black or Hispanic communities is historically and contemporarily different from that in the United States, and census data

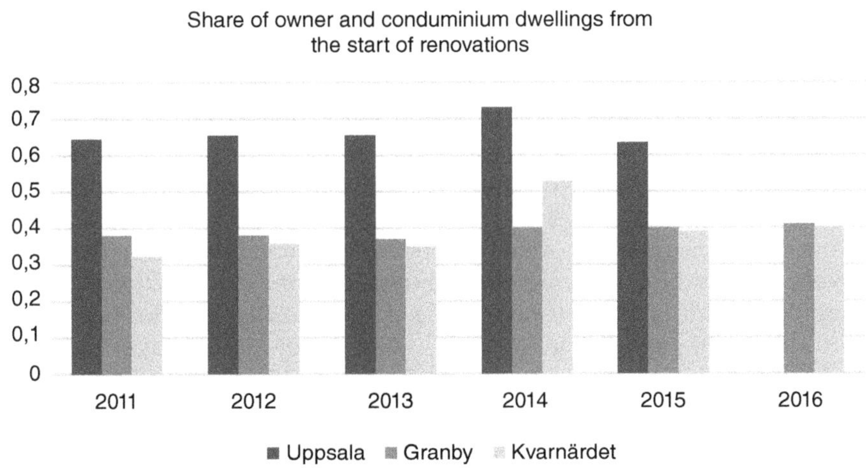

Figure 3.4 Tenure split in Uppsala and studied tracts.
Source: Data from the Department of Urban Management, Uppsala municipality.

in Sweden does not make distinctions based on ethnicity. However, census data looking at the share of persons with a foreign background (born outside of Sweden or with at least one parent born outside of Sweden) does show that the populations of both Kvarngärdet and Gränby are diversifying (see Figures 3.2 and 3.3). Again, this supports the case that both neighbourhoods are gentrifiable but not gentrified despite the rent gap having closed.

If not gentrification, then what? Towards accumulation in a post-gentrification landscape

Current renovation processes in Uppsala show a trend of displacement through various displacement pressures. Nevertheless, as outlined above, its neighbourhoods very much lack signs of gentrification. Most importantly, gentrifiers are seemingly absent. Income levels are stagnant in the neighbourhoods, and social unrest and petty crime are prevalent. Local media is rife with news articles on the problems and unrest, including the torching of cars by youth and the open sale of drugs. Some tenants are afraid:

> I only walk my dog during the day now. It's hard now in winter; I only walk around my own block a few times instead of wandering longer. I feel exposed. A bit afraid.
>
> (Jennny, interviewed in February 2018)

Nonetheless, profits are being extracted through the new rent 'corrections' that have entailed eighteen to sixty-five per cent rent increases for tenants, often without any discernible increases in standards of living. (On the contrary, the neighbourhoods and landlords were targeted by a television documentary in 2015 outlining and exposing the poor quality of the renovations.) The areas in question, along with plenty of others around Sweden (Baeten et al., 2017) are clearly closing the rent gap and extracting new increased land rents. Nonetheless, is what we are witnessing here really best labelled *gentrification*? Moreover, without affluent groups moving to the neighbourhoods in question, how are the new rent levels extracted? Below, I try to give an account for the former and outline a research agenda to answer the latter.

Answering the question of 'to be or not to be' for gentrification is simple. Whether or not you analyse gentrification from the side of capital movement (like Smith, 1979) or from the side of cultural change (like Ley, 2003), both approaches require a 'gentry' – gentry that either drives the shift through patterns of consumption or absorbs the increased cost of living driven by investment cycles. Thus far, Kvarngärdet and Gränby show no sign of such a gentry being present. Structural renovations close the rent gap quickly and efficiently, and evidently, a gentry is no longer needed for increased extraction of urban land rent. Instead, these costs are absorbed by the urban poor. This is not to say that this post-gentrification process is benign or friendlier to working-class tenants (in that their neighbourhoods remain intact). On the contrary, here the process is not the gradual shift that can be observed in some gentrifying cases and in the gentrification history of Sweden, but it is overtly violent (Baeten et al., 2017; Pull and Richard, 2019). Rent increases are immediate (even if landlords sometimes offer grace periods and step-wise rent increases over a few years), and the effects on disposable incomes and the everyday life of tenants are instantaneous. Furthermore, displacements are bountiful both in terms of tenant relocation in space (Boverket, 2014) and in terms of the destruction of place (Pull and Richard, 2019). The situation is a new one, and it seems to be something other than gentrification.

My respondents can no longer afford to buy bus tickets downtown. As one of my respondents puts it,

> I want to take my family on vacation, but even taking the bus to the library is an expense that has to be budgeted. You feel bad. You feel really bad.
>
> (Anna, interviewed in January 2016)

Their quality of life is diminishing as their low and decreasing disposable incomes restrict them to a life of immobility. Stories of squeezing together and sleeping in shifts are prevalent, and several tenants have concerns over their health due to monotonous, low quality food:

> I smoke, you know; that costs money, too much money. And the dog needs food. I have to go without food sometimes. Can't let the dog suffer and starve because I'm stupid enough to smoke. But it's not healthy eating poorly you know.
>
> (Amanda, interviewed in June 2016)

The economic pressure can indeed go far. One tenant, a retired social worker, recounts how the rent for her two-room, sixty-square-meter apartment costs her SEK 7,900 per month, which she pays from her SEK 11,900 pension. With hot water, television, and phone bills on top, it is hard to get the remaining money to cover other necessities. Others sell family heirlooms or take loans from friends. Thus, one answer to the question on new rent levels is that rent extraction is made through increased economic polarization. The disposable income levels of the poor are simply shrinking, but that is not the full picture of how rents are extracted. This is where more research is urgently needed – on how rent extraction and valorization of urban land is possible when tenants reach the absolute limits of survivability. Initial probing suggests that all municipal taxpayers are to an extent paying for the new land rent. The former chief of the municipal cultural department frankly admits in an interview (conducted in January 2015) that there have been cuts in their budget for cultural events and education as a direct response to the increased costs for housing subsidies and other economic subsidies. Numbers from the city development office show a growth of hundred per cent in social contracts in Gränby over the past three years. As this data does not cover a sufficiently long enough time span, and the increases only amount to a total of four per cent of the dwellings, no hard conclusions can be drawn. Nevertheless, the data points to one (out of many) possible mechanisms for rent extraction in a post-gentrification landscape – mechanisms we need to learn more about.

Conclusion

When Eric Clark met Neil Smith in Sweden, he had no gentrification battlefields to point to. However, he did see the signs of battlefields on the horizon. Now these battlefields appear to have manifested themselves, but they are perhaps not best captured by the concept of gentrification. We have 'conflicts of interest, displacement, personal tragedies, and the desperation that leads to battlefields' (Clark, 2005) – yes. But gentrification? Perhaps not. How best to conceptualize this increasingly violent production of urban space is still up for grabs. Here I suggest post-gentrification – not as a way to denote a complete rupture from the processes of gentrification but to suggest that this is, in many ways, something new. By post-gentrification I mean schemes aimed at increasing rent extraction through interventions in the built environment, but where the replacement of poor populations by a wealthier gentry is no longer as central, or at all necessary. Instead, the process of post-gentrification as it manifests itself in Uppsala is both displacing and emplacing populations. Some are forced to move, or to downsize to a smaller apartment within the neighbourhood, while others cannot move, from lack of affordable housing elsewhere in the municipality and region. Or put differently, mobility in the landscape of post-gentrification cannot be captured through a simple displacement-replacement dialectics where poor households are squeezed out and wealthier move in. The unifying experience of tenants can perhaps best be captured through displacement pressure. A pressure that is exerted whether it materializes in relocation by the household or not. Elsewhere, I and a colleague argued for a focus on dis*place*ment and domicide as concepts rather than gentrification (Pull and Richard, 2019), to forefront the changes to home and place for individual tenants that remain in these neighbourhoods. More research is undoubtedly needed on this new post-gentrification landscape, where poor populations are targeted, but in other ways than through their replacement by a gentry.

Notes

1 Social housing contracts are acute interventions by municipal authorities to prevent homelessness, whereby the municipality enters into a contract with social services and sublets to the household. Most of the social contracts are dispersed throughout the housing stock and between public and private housing companies (Boverket, 2014).

2 While rent-gap theory is an important part of production-side gentrification theory and subject to a scholarly debate itself (Clark, 1995; Slater, 2017), that debate and the nuances of land rent and rent gaps will have to be omitted in this short chapter.

References

Andersson, R. and Turner, L. M. (2014). Segregation, gentrification, and residualisation: From public housing to market-driven housing allocation in inner city Stockholm. *International Journal of Housing Policy*, 14(1), pp. 3–29.

Baeten, G. and Listerborn, C. (2015). Renewing urban renewal in Landskrona, Sweden: Pursuing displacement through housing policies. *Geografiska Annaler, Series B: Human Geography*, 97(3), pp. 249–261.

Baeten, G., Westin, S., Pull, E., and Molina, I. (2017). Pressure and violence: Housing renovation and displacement in Sweden. *Environment and Planning A*, 49(3), pp. 631–651.

Bostic, R. W. and Martin, R. W. (2003). Black home-owners as a gentrifying force? Neighbourhood dynamics in the context of minority home-ownership. *Urban Studies*, 40(12), pp. 2427–2449.

Boverket. (2014). *Flyttmonster till följd av omfattande renoveringar.* Landskrona: Boverket.

Clark, E. (1987). *The rent gap and urban change: Case studies in Malmo 1860–1985.* Lund: Lund University Press.

Clark, E. (1995). The rent gap re-examined. *Urban Studies*, 32(9), pp. 1489–1503.

Clark, E. (2005). The order and simplicity of gentrification: A political challenge. In: Atkinson, R. and Bridge, G. (eds.) *Gentrification in a global context: The new urban colonialism.* London: Routledge.

Freeman, L. (2005). Displacement or succession? Residential mobility in gentrifying neighborhoods. *Urban Affairs Review*, 40(4), pp. 463–491.

Grander, M. (2017). New public housing: A selective model disguised as universal? Implications of the market adaptation of Swedish public housing. *International Journal of Housing Policy*, 17(3), pp. 335–352.

Hajighasemi, A. (2004). *The transformation of the Swedish welfare system: Fact or fiction?* Huddinge: Södertörns högskola.

Hammel, D. J. and Wyly, E. K. (1996). A model for identifying gentrified areas with census data. *Urban Geography*, 17(3), pp. 248–268.

Hedin, K. (2010). *Gentrifiering, socialgeografisk polarisering och bostadspolitiskt skifte.* Phd dissertation, Lunds University, Lund.

Hedin, K., Clark, E., Lundholm, E., and Malmberg, G. (2012). Neoliberalization of housing in Sweden: Gentrification, filtering, and social polarization. *Annals of the Association of American Geographers*, 102(2), pp. 443–463.

Johansson, B. O. (1997). *Den stora stadsomvandlingen: erfarenheter från ett kulturmord.* Stockholm: Arbetsgruppen för arkitektur och formgivning, Regeringskansliet.

Ley, D. (2003). Artists, aestheticisation and the field of gentrification. *Urban Studies*, 40(12), pp. 2527–2544.

Marcuse, P. (1985). Gentrification, abandonment, and displacement: Connections, causes, and policy responses in New York City. *Washington University Journal of Urban and Contemporary Law*, 28, pp. 195–248.

McKinnish, T., Walsh, R., and Kirk White, T. (2010). Who gentrifies low-income neighborhoods? *Journal of Urban Economics*, 67(2), pp. 180–193.

Molina, I. (1997). *Stadens rasifiering: Etnisk boendesegregation i folkhemmet*. PhD dissertation, Uppsala University, Sweden.
Molina, I. and Westin, S. (2012). Renoviction - even in Sweden?: Four case studies on the regeneration of public housing estates. Paper presented at *AAG Annual Meeting of the Association of American Geographers*. Available at: http://uu.diva-portal.org/smash/record.jsf?pid=diva2%3A526369&dswid=-1641
Öresjö, E. (1993). *Att förnya bostadsförnyelsen*. Stockholm: Statens råd för byggnads forskning.
Öresjö, E. (1997). Lokalt områdesarbete i socialt utsatta stadsmiljöer: Exemplet Råslätt. In: Denvall, V. et al. (eds.) *Välfärdens operatörer: Social planering i brytningstid*. Umeå: Boréa, pp. 1859–1886.
Polanska, D. and Richard, Å. (2019). Narratives of a fractured trust in the Swedish model: Tenants' emotions of renovation. *Culture Unbound*, 11(1), pp. 141–164.
Pull, E. and Richard, Å. (2019). Domicide: Displacement and dispossessions in Uppsala, Sweden. *Social and Cultural Geography*, DOI: 10.1080/14649365.2019.1601245
Salonen, T. (2015). Allmännyttans skiftande roll på den lokala bostadsmarknaden. In: Salonen, T. (eds.) *Nyttan med Allmänyttan*. Interak: Liber, pp. 132–159.
Sandberg, L. A. (2014). Environmental gentrification in a post-industrial landscape: The case of the Limhamn quarry, Malmö, Sweden. *Local Environment*, 19(10), pp. 1068–1085.
Slater, T. (2017). Planetary rent gaps. *Antipode*, 49(1), pp. 114–137.
Smith, N. (1979). Toward a theory of gentrification: A back to the city movement by capital, not people. *Journal of the American Planning Association*, 45(4), pp. 538–548.
Thörn, C. and Holgersson, H. (2014). *Gentrifiering*. Lund: Studentlitteratur.
Thörn, C. and Holgersson, H. (2016). Revisiting the urban frontier through the case of New Kvillebäcken, Gothenburg. *City*, 20(5), pp. 663–684.
Westin, S. (2011). '... men vart ska ni då ta vägen?': Ombyggnation ur hyresgästernas perspektiv. Institutet för bostads- och urbanforskning, Uppsala universitet.
Wyly, E. Newman, K., Schafran, A., and Lee, E. (2010). Displacing New York. *Environment and Planning A*, 42(11), pp. 2602–2623.

4 Emotional displacement

Misrecognition, symbolic violence, and loss of place

Chiara Valli

I see the gut renovation my landlord did as soon as we moved out and feel disgusted. I see my best friends and families having to move to Cypress Hills and East New York and I am livid. I feel such an unending wave of anger and frustration. Sometimes I get off my train station on Morgan Avenue and feel suffocated by this rage when I see a sea of new faces and new businesses. I feel anger that the home that my family tried to carve out of this horrible little slice of earth called Bushwick is not mine anymore. Or rather, that the fruit now ripe for picking isn't ours.

(Rosa, 26, born in Bushwick to Dominican parents)

How we feel about another – or a group of others – is not simply a matter of individual impressions, or impressions that are created anew in the present. Rather, feelings rehearse associations that are already in place, in the way in which they 'read' the proximity of others, at the same time as they establish the 'truth' of the reading.

(Ahmed, 2004c: 39)

Introduction

Where do the feelings of disgust, fury, anger, frustration, suffocation, and rage expressed by Rosa in the opening quote come from? Specifically, what are the structural social relations that instigate feelings of displacement? And finally, what do these feelings *do*, and what *may* be done with them to resist displacement and gentrification?

As this edited book testifies, the issue of displacement in urban studies is central, yet still controversial. Some researchers insist that the actual displacement of residents in urban restructuring processes is a statistically marginal phenomenon (Vigdor et al., 2002; Freeman and Braconi, 2004); others sustain that the positive consequences of renewal and gentrification (in terms of increased services, amenities, safety, social mix, and public spaces improvement) are beneficial for local low-income groups who manage to stay put (Boddy, 2007; Freeman, 2011). Therefore, the displacement of low-income populations connected to urban renewal processes is a debated and questioned issue, and it is far from being fully understood and commonly accepted even in the academic field.

Moreover, pinpointing and quantifying displacement presents some serious methodological and conceptual challenges (Atkinson, 2002; Shaw and Hagemans, 2015). In fact, what is the most appropriate timeframe for measuring displacement? When exactly, in a process of continuous urban change, does displacement begin? And when does it end? When does the amount of displaced residents become 'statistically significant'? These are some of the possible questions on the definition and delimitation of the phenomenon of displacement that make its quantitative measurements a slippery and often not even fruitful approach.

On the other end, empirical qualitative studies of displacement, including the one presented in this chapter, show that, even before becoming 'statistically significant', displacement is a powerfully perceived, experienced, and dreaded phenomenon in people's everyday lives. The perceptions of the threats of displacement have real impacts on people's livability and affect their sense of home and belonging, their ways of relating to city spaces, and their stability and security (Slater, 2008; Davidson and Lees, 2010; Atkinson, 2015; Shaw and Hagemans, 2015; Baeten et al., 2017; Pull and Richard, 2019). Drawing on Ahmed (2004c) and Bourdieu (1984), I argue in this chapter that these perceptions are not simply personal idiosyncratic feelings but are embedded in and shaped by individuals' positioning within urban social structures. For critical researchers concerned with gentrification in its various shapes and forms, this should be reason enough to make displacement and the emotions and affects connected to it a primary field of inquiry. In this sense, the recently emerged interest in emotions and affects within sociologic and geographical research can add critical perspectives that question the conventional limits to displacement and gentrification scholarship. Moreover, emotional analysis is key also to research on protests and anti-gentrification struggles.

The present study on displacement draws on qualitative research on the effects and dynamics of gentrification in the neighbourhood of Bushwick, Brooklyn, New York City. The study pays attention to the empirical world, the lived experiences, and what displacement *feels like* in the everyday lives of residents who are still in place. Nevertheless, the goal is not a purely phenomenological exploration of residents' feelings; rather, as Jones and Jackson (2014: 15) put it, the goal is 'to consider where these feelings come from, what power relations they reflect, reinforce or undermine', with the conviction that '[t]he personal is the political; equally, the political is personal'. Ultimately, this study contributes to the definition of displacement by focusing on the intangible, yet vividly experienced, forms of gentrification-led 'loss of place'. Far from pertaining to the individual psychological sphere, affective and emotional components of displacement speak directly to urban sociological and geographical studies and therefore can critically nuance and complement more traditional class-, race-, and culture-based accounts of gentrification. Specifically, emotional displacement, as an effect of the class and racial violence inscribed in gentrification, is conveyed through what Bourdieu (1984) labels *symbolic violence* – that is, domination via various forms of misrecognition.

Affect and emotions of displacement in the gentrification literature

Arguably, the most productive approach within the 'affective turn' in the social sciences of the early 2000s is the critical strand of affective geography that focuses on demonstrating that 'struggles over place are often played out through emotional attachment' (Jones and Jackson, 2014: 1). This approach can critically and productively enrich our understanding of various forms of displacement. By focusing on the social and political constituents of emotions of displacement, this study originates in the conviction that 'social sciences must take as its object both this reality and the perception of this reality, the perspectives, the points of view which, by virtue of their position in objective social space, agents have on this reality' (Bourdieu, 1989: 18).

Early studies of gentrification-led displacement tended to approach the phenomenon in terms of material loss-of-space, that is, as 'the involuntary *dislocation of households from city neighborhoods* as more affluent households compete with them for the desirable older housing stock' (Hartman et al., 1982: 18, emphasis added). Nevertheless, although not explicitly formulated in terms of emotions and affect, some attention to displacement beyond physical relocation already appeared in early accounts of gentrification. For instance, what Marcuse almost thirty-five years ago formulated as 'pressure of displacement' is a potent and emotion-laden form of displacement derived from the shifted uses and demographics in a gentrifying neighbourhood, which precedes physical relocation:

> When a family sees the neighbourhood around it changing dramatically, when their friends are leaving the neighbourhood, when the stores they patronize are liquidating and new stores for other clientele are taking their places, and when changes in public facilities, in transportation patterns, and in support services all clearly are making the area less and less livable, then the pressure of displacement already is severe. Its actuality is only a matter of time. Families living under these circumstances may move as soon as they can, rather than wait for the inevitable; nonetheless they are displaced.
>
> (Marcuse, 1985: 207)

More recently, often drawing on feminist geographies and to the affective turn in the social sciences, the affective and emotional dimensions of displacement have gained a growing attention, pointing to the idea that people can experience displacement even when they are still in place and have not (or not yet) factually lost their homes.

Davidson (2009) suggests an understanding that emphasizes the notion of 'place' in dis*place*ment as 'lived experience of space'. Crucially, Davidson (2009: 228) argues that 'people can be displaced – unable to (re)construct place – without spatial dislocation, just as much as they can with spatial dislocation'. Inspired by Davidson, Stabrowski (2014: 796) presents 'everyday displacement' as 'the lived experience of ongoing loss of the security, agency, and freedom to "make place"' for lower-income residents in a gentrifying urban area. With a feminist framework and by drawing on theories of place attachment, Cahill (2007) uses the notion of 'cultural displacement': individuals' feelings of cultural belonging and security are developed in connection to the surrounding environment. Gentrification processes change this social and cultural context, and this can provoke a lost sense of identity and self, even when not connected to direct spatial displacement. Finally, Holgersson (2014) talks about 'discursive displacement', when she describes how the residents of a gentrifying area in Gothenburg, Sweden, get discursively (before physically) cut out from their neighbourhood in the official representations and narrations of ongoing redevelopment projects. Significantly, she underlines that discursive displacement together with a post-political 'win-win' rhetoric of gentrification as positive urban development 'undermine the possibility for people to politicize their exclusion' (118) and therefore displace residents *politically*, as much as materially.

Considered collectively, these and other contributions (Bondi, 1999; Kern, 2012; Shaw and Hagemans, 2015; Kern and McLean, 2017) point to the fact that intangible components of displacement – 'cultural', 'everyday', 'discursive', or 'emotional' and more – are significant and critical expressions of the power structures at play in urban renewal processes. As such, intangible and emotional dimensions of displacement have deep and longstanding roots in the social, political, and economic production of urban space.

A conceptual framework for emotional displacement: misrecognition, symbolic power, symbolic violence

In Valli (2015), I argued that interaction and encounter in the urban spaces of the neighbourhood are privileged moments to explore perceptions of gentrification and the emergence of emotional displacement. The encounter with others brings to surface emotions that rehearse relations and associations that are already in place and embedded in social power structures (Ahmed, 2004b). Here, I will argue that emotional displacement originates from what Bourdieu (1979, 1984, 1989) named *misrecognition* and *symbolic violence*.

Symbolic power consists of tacit – and largely unconscious – modes of social and cultural domination that reinforce individuals' placement in social hierarchies through everyday practices, habits, institutions, and behaviours. Symbolic power is the power of acknowledging, valuating, and consecrating:

> Symbolic power is the power to make things with words. ... [It] is a power of consecration or revelation, the power to consecrate or to reveal things that are already there. ... a group, a class, a gender, a region or a nation begins to exist as such, for those who belong to it as well as for the others, only when it is distinguished, according to one principle or another, from other groups, that is, through knowledge and recognition.
>
> (Bourdieu, 1989: 23)

Hence, symbolic power sustains and reproduces itself via the devaluation and *misrecognition* of objects, entities, power relations, and subjects that exist but become deprived of their meaning in order to legitimate the meanings and values attributed by the dominating groups. Misrecognition is the medium through which to exercise *symbolic violence*. Symbolic violence consists of 'the imposition on subordinated groups by the dominant class of an ideology which legitimates and naturalizes the status quo' (*Oxford Reference*). Importantly, symbolic violence is usually understood as exercised upon a social agent with his or her (although mostly unconscious) complicity, gender relations being a paradigmatic example (Bourdieu, 2001). Symbolic power is so profoundly fastened in social practices that it is hardly acknowledged and therefore is reproduced with the complicity of the dominated group in ambiguous power relations. Since it resonates with individuals' education, habits, and cultural expectations, symbolic violence is hardly called into question.

This chapter proposes a reading of gentrification as violence and, in particular, displacement as originating in economic, racial, as well as symbolic violence via misrecognition. However, the presented empirical materials depart from and challenge the deterministic view of dominated groups as simply unconscious, complicit agents in the reproduction of power relations. As I will illustrate briefly in the coming empirical section, emotional displacement can take the form of rage and outrage. These feelings are interpreted here in an emancipatory way, as a potential powerful trigger for resistance to displacement.

The coming section reports two stories from long-time Bushwick residents as empirical starting points to illustrate two specific kinds of misrecognition. The first kind of misrecognition connects to and expands Marcuse's (1985) definition of *displacement pressures*. The latter relates to symbolic capital displays in the neighbourhood spaces and the kinds of conflict originating in different habitus and capitals in gentrifying urban spaces. Hence, the coming empirical section has two aims. First, it presents and unpacks the emotional responses of two long-time residents in a gentrifying neighbourhood. Second, it reflects upon the power relations these feelings are coming from, react to, challenge, or reinforce. Before the analysis, some background information on the study case and methodology for investigation are laid out.

Stories of emotional displacement

Bushwick is a predominantly Latino neighbourhood located in North-West Brooklyn, New York City. Today, the largest share of the population is Hispanic or Latino, accounting for 65.3 per cent of the population. The second largest group is Black or African American, with 17.3 per cent, followed by 10.5 per cent White, and 4.6 per cent Asian. Bushwick is listed amongst the most rapidly and intensely gentrifying areas in New York City, with an over forty-four per cent increase in rent between 2000 and 2014, against a nineteen per cent increase city average (NYC Furman Center 2016). The area is undergoing rapid ethnic change as well. Between 2000 and 2014, the number of White residents showed a 322 per cent increase (from about 3,000 to nearly 13,000), accompanied by a general increase in young adult households (18–35 years old) and households without children (Valli, 2017). The first signs of gentrification became apparent in the neighbourhood spaces around the mid-2000s. Just some years earlier, artists started converting and populating the stock of vacant warehouses in the North West area, mostly pushed out by the hyper-gentrification of the neighbouring and 'ultra-hip' Williamsburg. The influx of artistic activities was swiftly followed by a bourgeoning consumption scene, which made Bushwick reach *Vogue*'s top ten 'coolest neighborhoods in the world'. Following a well-rehearsed pattern in the city, predatory real estate practices followed, and property values increased outstandingly (Zukin and Braslow, 2011).

The empirical materials presented here were collected as part of a broader research project (Valli, 2015, 2017) conducted in 2013–2014 in Bushwick. About forty semi-structured in-depth interviews with long-time and new residents were conducted for the study; they were complemented with secondary data from planning documents, media print, social media, and research dissemination activities.

In Valli (2015), I showed that the encounter with bodies that hold more privileged class and racial positions in the spaces of the neighbourhood that previously felt like home erode long-time residents' sense of belonging and ability to take and make space. These feelings hold fundamentally political value because they are constituted in and of objective social structures and power positions. In fact, as widely argued by feminist researchers, social class and its intersections with gender and race 'are simultaneously subjective, structural, and about social positioning and everyday practices' (Brah and Phoenix, 2004: 80). With these premises, my research in Bushwick (Valli, 2015) showed that long-time residents feel 'othered' by the newcomers, and the hidden and overt power relations at play prevent them (directly or indirectly) from frequenting some neighbourhood spaces. Their ability to take and make place is surpassed by that of the newcomers. Moreover, long-time residents assume a defensive position against the neighbourhood changes by 'othering' and dis-identifying themselves from the newcomers, stereotypically labelled as 'hipsters', in an attempt to reinforce their claims of authenticity and belonging in the neighbourhood. Importantly, for long-time residents, newcomers are living embodied reminders of structural inequalities and unbalanced power relations. With gentrification, they feel they are being deprived of their sense of place, on top of compelling threats of material displacement.

I hereby draw and extend those findings by diving deeper into two specific stories from my interviews with long-time Bushwick residents. The stories will help highlight different forms of symbolic violence and misrecognition that lie at the core of emotional displacement as it is perceived by long-time residents in gentrifying Bushwick. Importantly, I suggest here that emotional displacement can become a trigger for resistance and a seed for empowerment.

Rosa, 26, student, born in Bushwick with Dominican background

> I lived between Halsey and Hancock, which were two really bad streets back in the days, and they were so bad that I had to take the long way home to go back from elementary school; if you wanted to go to the corner store, you had to go around to avoid certain places. That neighbourhood was bad because there were a lot of drug dealers, gangs in the area, and a high police presence. My parents and a couple of other people in the block started a block association, and they collected money, and we had these awesome block parties. All the houses were competing with each other to make the best Christmas decorations and Halloween decorations. Those are the kind of things that make a community. People always watched out for each other. Every time something was going on, that particular block was very united. Like, during the 2003 black-out, my mum made this giant pot of *asopao* and she fed the whole block that day.
>
> So, what I struggle with is that when people talk about new people coming in and when they say it in a positive light, it's almost like as they are 'cleaning it up'. When it's always been pretty good. Where I lived, people worked hard to make their neighbourhood great and they did it. It wasn't safe in certain areas, but wherever we called 'home', we made a home out of it. When I got to my block, I always felt safe.
>
> It makes me really angry when people say that the people who lived previously in now gentrified areas didn't care about where they lived. If they didn't care about where they lived, would they be living like fucking animals? No! If you come on a morning on a Saturday or a Sunday, you will see people sweeping the streets; it's because they are taking care

> of their stuff. If you see their plants hanging up against the windows, it's because they are taking care of their stuff. If you live in a place, you are not going to leave it go to shit, right? But of course if you don't have the means to pay for remodel, you are not going to do it. [...]
>
> A lot of people have 'made Bushwick precious' and fetishized it and think of it as something really nice. But this is just people's home. This is fucking home. This is what it is. It's not an art project. [...]
>
> In a neighbourhood that is gentrifying and changing so much, every interaction is loaded. Everything is heavily weighted, because the stakes are just so high. I'm losing my home because I don't make enough to live here. Because there are people who looked different than me, who have more money, and supposedly put more value just by *who* they are. What I am is not valuable enough. I don't mean shit. That's what makes people so angry. I don't mean anything; my likes and what I think don't have a place at the table here anymore, even if I've been here when it was so shitty and tried to make something of it.

The quote opens with a description of Rosa's personal experience of growing up in a 'tough' area of Bushwick, with a high rate of criminal activities related to drug dealing. She focuses on how hers and other families from her street built a social network of support and created a sense of community that helped them deal with the dangers associated with the presence of nearby criminal activity and the sense of insecurity caused by a high police presence. She expresses a great sense of pride about how her family and other neighbouring families, despite the difficult circumstances and limited economic resources, created a hospitable home for themselves in the neighbourhood. Hence, she feels that the trope of gentrification's alleged effect of 'cleaning-up' a dangerous area from crime and blight is misplaced and unfair because it misrecognizes the efforts of the people who have struggled to make a home there, long before the gentrification process started.

Moreover, Rosa warns about the process of 'place fetishization' associated with gentrification. The symbolic value attached to gentrifying areas as 'hip' or 'cool' or 'up and coming' is an expression of the gentrifiers' symbolic power 'of consecration or revelation of ... things that are already there' (Bourdieu, 1989: 23). What was already there, in this case, is Bushwick as somebody's *home*. The fact that some groups of people (gentrifiers, real estate, media) have the power of re-consecrating those places through new values and meanings ('making it precious') implies that they also hold the power of devaluating and misrecognizing what and whom had already been there for a long time. This is a form of symbolic violence inscribed in class- and race inequality typical of gentrification's place-making modality – conveyed by the presence of new residents who allegedly 'put more value just by *who* they are' because they hold more prestigious forms of economic, cultural, and ethnic capitals.

Rosa's feelings exemplify the connection between symbolic violence and emotional displacement. The sense of displacement, articulated by Rosa as feelings of anger and frustration, comes from the devaluation and misrecognition process associated with the influx of new, wealthier groups and their symbolic power. This symbolic domination causes a literal and metaphorical loss of place for lower-income residents and is therefore at the origin of their displacement.

Carlos, 36, teacher, born in Bushwick with Puerto Rican background

CV: What about all the artistic activities that are going on? Do people you know go there and enjoy them?

There were a lot of high points where people were enjoying the beautification, maybe some places that were littered, dirty, they cleaned it up. Art is art and is beautiful, but then some people take a nice platform and then they might abuse the opportunity and put up things that might seem offensive to the people who grew up there.

If you go just around the corner, to the *** gallery, you notice that they have made a big mural on the side of their building, and people don't understand. It looks like a portrait of Jesus Christ, with a halo around his head, holding a lamb, a famous picture ... Well, they have the same picture but with what looks like an animal; it looks like a cartoon of a turd. The local religious community took great offense to that. Because when you walk around, it looks like a mockery to their personal faith. And it's huge!

C: Did they say anything to the gallerist?

I think people don't know who to say something to. But when I speak to people in the community, they are not happy about it. When I walked by and saw the murals, I didn't know anything of it. At first I was offended, because of my personal belief, and then a part of me who doesn't want to be cynical or critical says 'maybe that's some artists' portrayal of something ...', but ... it grieved me. If you do art, to me, I deem you as an irresponsible or negligent artist if you express art that offends people who live in the community. So, now I question your art: What's more important, your art or the community? Are you a person who is looking to do art that expresses a message that looks to offend? Are you being political with your art? What is your agenda? What is your purpose? What is your intention? The absence of information only allows people to speculate. If you are going to express a message, and you are not going to make information available, you shouldn't be mad when people speculate, because that's their right to speculate. Especially when it may offend them. I don't know what the murals is about [...] but the everyday person who lives in this community may see that and they may feel that their personal belief are being mocked.

You are doing art, but art that is catering to a targeted group of people, whether they are artists or people of a certain belief, like atheism; when you walk by, people can interpret that work of art differently. It depends on the lens with which they look. An artist can be like 'wow, that

> is fascinating!' To the kid in the community whose mother is a devoted catholic, or his parents are Pentecostal, I don't know, they might look at them and say 'who are these new people moving to our neighbourhood and making fun of my faith?' These are unintended consequences. And for me, unintended consequences are there when you don't think things through, or the truth is maybe they are intended consequences, and if they are intended consequences, it means you thought it through and it wasn't really important to you whatever the consequences were going to be. Because what your agenda was, superseded what you knew the reactions were going to be.

Carlos' story demonstrates that displacement feelings are not only triggered by the fear of losing one's home, or by the progressive unaffordability of consumption places (i.e., the pressures of displacement defined by Marcuse). Conflicts and feelings of misrecognition arise also over the display of symbolic capital through aesthetic and artistic disposition.

Carlos explained that a mural on the wall of a new art gallery in the neighbourhood had triggered feelings of indignation and offense amongst many of his acquaintances who frequent a nearby catholic church. The mural was interpreted by many locals as a caricature of Jesus and perceived as a mockery of the catholic religious faith, widespread among the long-time population in Bushwick. A week before my interview with Carlos, I had actually interviewed the owner of the *** gallery, a young female artist from Long Island who had opened her gallery about a year earlier and envisioned it as 'an alternative art space and community outreach venue'. Paradoxically, in our talk, she affirmed that the main aim of the gallery was 'to improve community' and that through a variegated offer spanning from exhibitions to art classes, she was 'trying to include as many people as possible'. She explained,

> Connecting to local people is a difficult task. That's an issue in itself: the lack of communication between cultures. A lot of people here don't have e-mails, or phones, so you need someone to build trust and bridge that gap. Another thing is the language barrier with the Hispanic community, because, for example, the kids I work with speak English very well, but their parents don't always do that. So, to get the parents involved in what their kids are doing is a really big struggle. And I have been sending flyers in English and Spanish, but it's still hard Parents still don't come; they are sceptical. Especially if you have a new business, people don't trust you. It's hard. [...] The only way to beautify without gentrifying in my opinion is through education. Educating people to art, to their rights as renters ... and that is what we are trying to do.
>
> (Nora, 29, owner of the ***gallery)

Although in good faith, Nora problematically constructs local residents through a *deficit* lens: her intentions are to 'reach out to', 'include', 'improve', 'educate', and 'involve' the 'local communities'. She blames her struggles to achieve those goals on the fact that, in her opinion, many locals *do not* speak English, *do not* use e-mails, *do not* trust her or other new businesses. When I asked if she had perceived any conflicts when she opened her gallery there, she said that she did not observe anything negative and that people were treating her in a welcoming way.

By considering Carlos's and Nora's perspectives together, it is possible to make some observations about the sociocultural and communicative divide between Bushwick's long-time and new residents and about the individual ethical aspects of the gentrifiers' actions, their symbolic power, and its consequences in terms of symbolic and emotional displacement of long-time residents.

Gentrification as class war over space (Smith, 1996) and its inscribed symbolic violence do not happen intentionally through the action of individuals. Nevertheless, as Wilkinson (2016) put it, 'while the wider structures of gentrification may exist beyond the agency of individual retailers and consumers, this does not mean that individuals have no role to play in determining how gentrification plays out in our communities'. Bourdieu explains that taste functions politically as a vehicle of domination in class conflict. Misrecognition passes through internalized taste (habitus) and, hence, might be unintentional, but it is still a form of domination. Thus, the case of ambivalent religion-themed murals in front of a local church in Bushwick serves as another example of the relationship between symbolic violence and displacement. Liberal values such as freedom of expression are prone to become vehicles of domination and reaffirmation of privilege if not calibrated vis-à-vis social concerns. Such violence may (1) trigger emotional displacement and (2) echo and amplify structural forms of privilege and domination.

Especially in a hastily gentrifying context, the display of symbols and aesthetics is power-laden and a vehicle of contrasting modes of place-making. Accordingly, the symbolic violence that comes from misrecognition is not less political or violent than physical displacement, as it is an expression and tool of class- and racial domination. The aesthetics of new residents, businesses, and lifestyles become highly symbolic of wider forms of displacement and class violence.

Conclusions

This chapter opened by stating the contested meaning and significance of displacement in contemporary gentrification studies. Paraphrasing Bondi and Davidson's (2011) argument that the definitions of affect and emotions, in order to be useful and productive, need to be used relationally and openly, I argue that displacement, too, as being fundamentally characterized by emotions and affects, needs to be interpreted as a theoretical concept in-use and in-the-making.

To attend to that, in this chapter, I combined literature on displacement with readings of Ahmed and Bourdieu. This allowed me to unpack some feelings of gentrification-led displacement and then re-embed them into structural power relations. The core argument is that emotional displacement derives from the symbolic violence of misrecognition. However, departing from over deterministic structuralist perspectives, this contribution also highlights the empowering potential for resistance intrinsic in emotional displacement. In fact, the other side of the violence coin is the possibility for resistance. As the quotes presented here testify, misrecognition can provoke emotional displacement in the form of frustration, outrage, and rage. These feelings can, in turn, become the first trigger for resistance.

In the face of landlord abuses, police harassment, racial discrimination, real estate predatory speculations, and eviction threats (Betancur, 2011; Paton and Cooper, 2016), on top of the described 'softer' sides of violence, how can the articulation of emotional displacement inform anti-gentrification struggles against *material* displacement? Recalling Ahmed (2004a), what can emotions *do* in the struggle to stay put?

In some contexts, gentrification's symbolic violence triggered feelings that were so forceful that they resulted in violent anti-gentrification protests. For example, in 2015 on Brick Lane, East London, individual retailers were singled out and attacked as exemplifying the cultural domination attitude (Wilkinson, 2016). In other contexts, emotional displacement triggered forms of emancipatory reactions and resistance. For instance, Sandoval (2018: 10) shows that gentrification pressures in three Latino *barrios* in US cities encouraged the re-emergence of neighbourhood-based Chicano arts and Latino-owned local businesses 'as young artists are taking up the mantle and using art to transform the neighbourhood, but in their own terms'.

Ultimately, research into emotional displacement can inform anti-gentrification struggles in several ways. First of all, for long-time residents who find themselves feeling angry and vulnerable in the face of the misrecognition of their existences and cultural values, and who even face risks of material displacement, formal accounts of emotional displacement can work as prompts that such feelings are not subjective, idiosyncratic, or experienced in isolation. They are shared and expressions of the class war that gentrification represents. For new residents and businesses that come – by choice or constrains or, more likely, a combination of the two (for a self-critical account, see Schlichtman et al., 2017) – to establish themselves at the frontline of gentrification, research on emotional displacement can suggest the importance of practicing an ethics of care, which would imply at the very least to acknowledge the existence, the value, and the struggles of the long-standing residents.

In this sense, it seems appropriate to recall Tom Angotti's call for actions about things that *can* be done to resist gentrification, where reciprocal acknowledgement and communication are at the basis of possible organized resistance:

> TALK TO PEOPLE on your block and in your neighborhood. If you are a gentrifier, talk to the people getting gentrified. If gentrifiers are moving in and you're afraid you're going to be pushed out, talk to the gentrifiers. If we don't talk with each other we can't work together and if we don't work together gentrification will proceed unchecked. Talk, argue, become friends or enemies, but if we ignore each other it will only deepen the divide and make it harder to change anything. These connections are the best foundation for building alliances and creating community coalitions that can stop displacement and protect residents and businesses. Too many people get outraged at what's going on and never talk to anyone else about it, least of all the people who are 'different'.

Finally, greater and systematic attention to emotional displacement would be productive for gentrification researchers as well. I would like to welcome Rosa's hint above, about the risk, not only for new residents in gentrifying areas but also for well-intentioned researchers, of fetishizing gentrifying neighbourhoods. A deeper attention to the emotions and feelings of displacement in urban research would help us remember that 'this is just people's home ... It's not an art [or research] project'. Keeping that in mind would help make displacement research more self-reflective, self-critical, and hence more poignant.

References

Ahmed, S. (2004a). Affective economies. *Social Text*, 22, pp. 117–139.

Ahmed, S. (2004b). Collective feelings: Or, the impressions left by others. *Theory, Culture and Society*, 21, pp. 25–42.

Ahmed, S. (2004c). *The cultural politics of emotion*. Edinburgh: Edinburgh University Press.

Atkinson, R. (2002). *Does gentrification help or harm urban neighbourhoods? An assessment of the evidence-base in the context of new urban agenda*. ESRC Centre for Neighbourhood Research Glasgow.

Atkinson, R. (2015). Losing one's place: Narratives of neighbourhood change, market injustice and symbolic displacement. *Housing, Theory and Society*, 32, pp. 373–388.

Baeten, G., Westin, S., Pull, E., and Molina, I. (2017). Pressure and violence: Housing renovation and displacement in Sweden. *Environment and Planning A: Economy and Space*, 49(3), pp. 631–651.

Betancur, J. (2011). Gentrification and community fabric in Chicago. *Urban Studies*, 48, pp. 383–406.

Boddy, M. (2007). Designer neighbourhoods: New-build residential development in nonmetropolitan UK cities – the case of Bristol. *Environment and Planning A*, 39, pp. 86–105.

Bondi, L. (1999). Gender, class, and gentrification: Enriching the debate. *Environment and Planning D: Society and Space*, 17, pp. 261–282.

Bondi, L. and Davidson, J. (2011). Lost in translation. *Transactions of the Institute of British Geographers*, 36, pp. 595–598.

Bourdieu, P. (1979). Symbolic power. *Critique of Anthropology*, 4, pp. 77–85.

Bourdieu, P. (1984). *Distinction: A social critique of the judgement of taste.* Cambridge, MA: Harvard University Press.

Bourdieu, P. (1989). Social space and symbolic power. *Sociological Theory*, 7, pp. 14–25.

Bourdieu, P. (2001). *Masculine domination.* Stanford, CA: Stanford University Press.

Cahill, C. (2007). Negotiating grit and glamour: Young women of color and the gentrification of the Lower East Side. *City and Society*, 19, pp. 202–231.

Davidson, M. (2009). Displacement, space and dwelling: Placing gentrification debate. *Ethics, Place and Environment*, 12, pp. 219–234.

Davidson, M. and Lees, L. (2010). New-build gentrification: Its histories, trajectories, and critical geographies. *Population, Space and Place*, 16, pp. 395–411.

Freeman, L. (2011). *There goes the hood: Views of gentrification from the ground up.* Philadelphia, PA: Temple University Press.

Freeman, L. and Braconi, F. (2004). Gentrification and displacement New York City in the 1990s. *Journal of the American Planning Association*, 70, pp. 39–52.

Hartman, C. W., Keating, W. D., and LeGates, R. T. (1982). *Displacement, how to fight it.* Natl Housing Law Project.

Holgersson, H. (2014). Post-political narratives and emotions. In: Jones, H. and Jackson, E. (eds.) *Stories of cosmopolitan belonging: Emotion and location.* London: Routledge.

Jones, H. and Jackson, E. (2014). *Stories of cosmopolitan belonging: Emotion and location.* Hoboken, NJ: Routledge.

Kern, L. (2012). Connecting embodiment, emotion and gentrification: An exploration through the practice of yoga in Toronto. *Emotion, Space and Society*, 5, pp. 27–35.

Kern, L. and McLean, H. (2017). Undecidability and the urban: Feminist pathways through urban political economy. *ACME: An International E-Journal for Critical Geographies*, 16, pp. 405–426.

Marcuse, P. (1985). Gentrification, abandonment, and displacement: Connections, causes, and policy responses in New York City. *Washington University Journal of Urban and Contemporary Law*, 28, p. 195.

NYC Furman Center. (2016). *State of New York City's housing and neighborhoods in 2015. Focus on gentrification.* New York: NYU Furman Center.

Oxford Reference. Symbolic violence. Available at: https://www.oxfordreference.com/view/10.1093/oi/authority.20110803100546777 [Accessed 2020].

Paton, K. and Cooper, V. (2016). It's the state, stupid: 21st gentrification and state-led evictions. *Sociological Research Online*, 21, pp. 1–7.

Pull, E. and Richard, Å. (2019). Domicide: Displacement and dispossessions in Uppsala, Sweden. *Social and Cultural Geography*, pp. 1–20.

Sandoval, G. F. (2018). Planning the barrio: Ethnic identity and struggles over transit-oriented, development-induced gentrification. *Journal of Planning Education and Research.* doi: 0739456X18793714

Schlichtman, J. J., Patch, J., and Hill M. L. (2017). *Gentrifier.* Toronto: University of Toronto Press.

Shaw, K. S. and Hagemans, I. W. (2015) 'Gentrification without displacement' and the consequent loss of place: The effects of class transition on low-income residents of secure housing in gentrifying areas. *International Journal of Urban and Regional Research*, 39, pp. 323–341.
Slater, T. (2008). 'A literal necessity to be re-placed': A rejoinder to the gentrification debate. *International Journal of Urban and Regional Research*, 32, pp. 212–223.
Smith, N. (1996). *The new urban frontier: Gentrification and the revanchist city*. London: Routledge.
Stabrowski, F. (2014). New-build gentrification and the everyday displacement of Polish immigrant tenants in Greenpoint, Brooklyn. *Antipode*, 46, pp. 794–815.
Valli, C. (2015). A sense of displacement: Long-time residents' feelings of displacement in gentrifying Bushwick, New York. *International Journal of Urban and Regional Research*, 39, pp. 1191–1208.
Valli, C. (2017). *Pushing borders: Cultural workers in the restructuring of post-industrial cities*. Doctoral dissertation, Department of Social and Economic Geography, Uppsala University.
Vigdor, J. L., Massey, D. S., and Rivlin, A.M. (2002). Does gentrification harm the poor? [With comments]. *Brookings-Wharton papers on urban affairs*, Brookings Institution Press, pp. 133–182, http://www.jstor.com/stable/25067387.
Wilkinson, E. (2016). 'Let us devastate the avenues where the wealthy live': Resisting gentrification in the 21st century city. *Sociological Research Online*, 21, pp. 1–7.
Zukin, S. and Braslow, L. (2011). The life cycle of New York's creative districts: Reflections on the unanticipated consequences of unplanned cultural zones. *City, Culture and Society*, 2, pp. 131–140.

5 The biopolitics of displacement

Komfortfokozat, racism, and post-socialist gentrification in Budapest's Eighth District

Jonathan McCombs

Introduction

In early 1992, Márta Blaha published an article titled 'Evictions' in the *Amaro Drom* magazine, a publication of Roma politics and culture, speculating on the possibility of housing justice after the post-socialist turn. Juxtaposing a series of pictures of Roma families outside of their houses with their belongings stacked on the street, Blaha (1992) commented on the swiftness with which authorities moved their furniture out of the apartment, removed doors and windows, and dismantled the kitchen and bathroom, rendering the space uninhabitable. A little over twenty-five years later, the Mayor of Budapest's Eighth District, Mate Kocsis, signed an agreement to secure two billion forints, roughly seven million US dollars, to begin a project billed as an 'urban development complex' (Ványi, 2018). As the weekly district magazine reported, these new initiatives are not just about 'the facades, the streets, and the common spaces, but also the residents themselves' (Ványi, 2018). The article listed vocational training, economic enhancement, enterprise development, crime prevention, drug reform programs, and remedial education as the primary social programs targeting residents (ibid). Both these articles exhibit the biopolitics of gentrification in Budapest's Eighth District. Blaha's article describes how housing is a crucial node in *life processes*, where the state's ability to render a flat or house uninhabitable arbitrates who is able to maintain access to the implements necessary to sustain life in the city. The second article outlines how the Eighth District's development initiatives rely on strategies that intervene at the level of the population. In the case of the Eighth District, which has the highest population of Roma in the inner-city, I show how these population-management strategies are racialized.

I argue in this chapter that a biopolitical framework allows us to incorporate undertheorized complexities of property ownership patterns present in Budapest, and post-socialist cities in general. The process of privatization created a 'super-homeownership' property regime (Stephens et al., 2016), where the majority of the Hungarian population owns property outright. This property regime limits the impact of the market on gentrification processes because there is not a concentration of property in the hands of a landlord and banking class that can induce displacement. As such established gentrification theories (like the rent-gap theory) fall short of incorporating these complexities in part because they were developed in an Anglo-American context, where a landlord and banking class holds liens or title over a majority of housing. A biopolitical framing allows us to account for gentrification processes while considering post-socialist property regimes. To operationalize biopolitics in this case, I show how gentrification has been driven by interventions into the urban fabric through knowledge gleaned via two statistical categories: *komfortfokozat* (or comfort level) and Roma as a subset of the Eighth District population.

This book chapter is based on three stints of field work in Budapest's Eighth District over the summers of 2016 and 2017, as well as a year-long stint from August 2018 to July 2019. Over the course of this fieldwork, I conducted thirty interviews with activists, policy makers, and residents, as well as collecting thousands of pages of archival data from various archives across Budapest. Some statistical data was also collected as a part of this fieldwork. It should be noted that at the time of writing, the analysis of each of these sources is still ongoing.

Rethinking post-socialist gentrification: super-homeownership and biopolitics

Gentrification is the process of tenure change in cities, where higher-income residents replace low-income residents accompanied by, or as a result of, increased investment into the built environment (Lees et al., 2016: 9). Within urban studies, there is a consensus among Marxist scholars that gentrification is an integral part of the political economy of cities under capitalism, and with the expansion of capitalism, gentrification is globalizing as well (Lees et al., 2016). Scholars working within post-colonial and comparative urbanist frameworks have contested the universality of gentrification, citing among other things the heterogeneity of property regimes across the globe (Ghertner, 2015). As a result, comparative urbanists have urged context-specific approaches that highlight the histories and institutional configurations of particular cities (Robinson, 2006; Roy, 2009; Ghertner, 2015; Derickson, 2016). Within post-socialist gentrification studies, scholars working within this tradition have questioned the utility of Marxist gentrification theory for post-socialist cities, citing different institutional and political histories emanating from the socialist transition (Weist, 2012; Bernt, 2016; Ferencuhova, 2016; Tuvikene, 2016). Nevertheless, no consensus about what constitutes post-socialist gentrification has emerged.

Following Stephens et al. (2016), I argue that post-socialist countries have a specific property structure that emerged as a result of privatization (Marcuse, 1996). While there is considerable variation between each country, the net result of these privatization schemes led to high rates of owner-occupancy, the legacy of which remains almost thirty years after the transition (OECD, 2016).

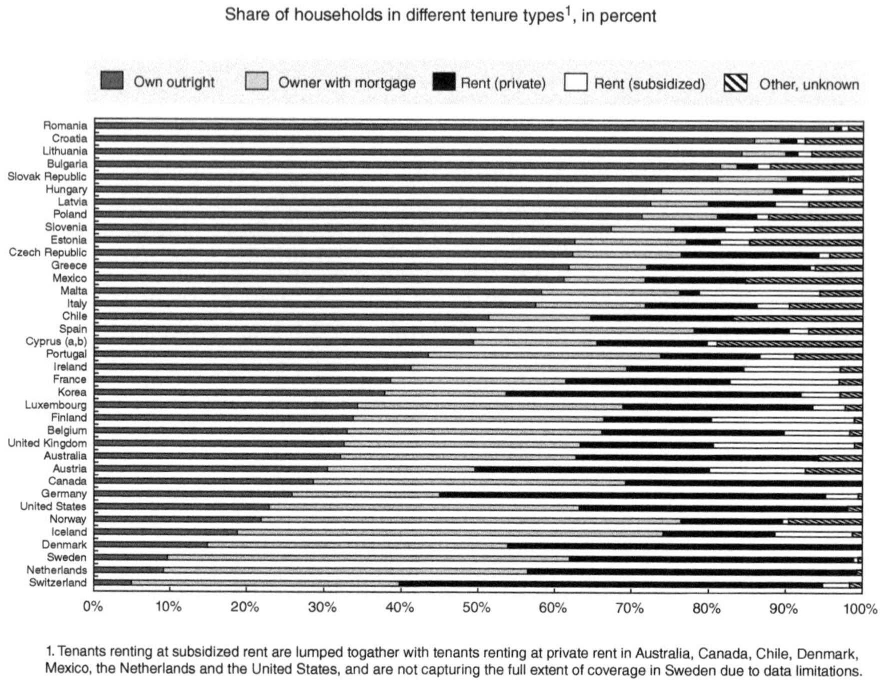

Figure 5.1 Housing tenure distribution, 2014, or latest year available.
Source: OECD (2016).

Every post-socialist country has an owner-occupancy rate of at least eighty per cent, and an outright owner occupancy rate (without liens) of at least sixty per cent (Stephens et al., 2016: 1221–1222). In countries with a legacy of capitalist housing regimes, owner-occupancy averages around sixty per cent of the housing market, with roughly half of owner-occupied housing owned outright. The UK and North America have outright owner occupancy rates between twenty and thirty per cent. Stephens et al. (2016) argue that post-socialist countries adhere to a 'super-homeownership' housing model that emerged from the welfarist tendencies of the former state-socialist regime.

Hungary exhibits the same tendencies as other post-socialist cities, with an outright owner-occupancy rate of seventy-five per cent and an overall owner-occupancy rate of almost ninety per cent (OECD, 2016). With such high rates of outright owner-occupancy, the displacement pressure from landlords or banks is lessened compared to homeownership regimes where property is concentrated within a landlord, developer class (Harvey, 1974). Theories of gentrification in Hungary have so far not addressed the diffuseness of homeownership – instead relying on variations of Anglo-American theories, such as rent-gap (Smith, 1996) and state-led gentrification (Hackworth and Smith, 2001), to explain phenomenon of displacement (Nagy and

Timár, 2012; Kovács et al., 2013; Czirfusz et al., 2015; Fabula et al., 2017). While these studies have provided valuable insights into gentrification processes in Hungary, few have meaningfully drawn attention to the spatial distribution of property ownership specifically as a key factor in post-socialist gentrification.

In Hungary, the process of privatization resulted in a patchwork of privately owned flats throughout the city as flats were sold as condominiums, leaving common spaces to be managed cooperatively – those flats that were not purchased by the occupiers remained under the control of the state (Hegedűs et al., 1996; Bodnár, 2004). In Budapest, the district was granted rights in 1990 to privatize housing as well as administer public housing (Bodnár, 2004), which meant that a single block of flats could have a mix of private and district-owned individual flats. Consequently, any given building throughout the city was likely to have a multitude of actors with different interests and priorities, making wholesale acquisition of buildings difficult. Bernt (2016) sums up these complexities through a case study of Saint Petersburg:

> Within the same building, [real-estate speculators] might thus meet owner-occupiers who can hardly make a living and who rent out parts of their apartments in order to obtain additional income; private tenants who can easily be displaced when a deal is set with the absent owner of the apartment; municipal tenants who need to be resettled by the city; owner occupiers who have just bought their apartment and are highly indebted with mortgage payments; and outright owners who want to stay put, as well as better off households who wait for a chance to improve their housing situation.
>
> (Bernt, 2016: 572–573)

Such a housing regime resulted in multiple, often conflicting property claims on any given building (see also 'anti-commons' literature, Heller, 1996; Buckley and Mathema, 2017).

In this chapter, I argue that within a 'super-homeownership' housing regime, the state, and its normative functions, plays a substantial role in gentrification and displacement processes. In this vein, I employ a biopolitical framework to theorize processes of gentrification and displacement in the Eighth District. Rabinow and Rose (2006) understand Foucault's conception of biopower (see Foucault, 1976) as concerning the tactics for managing the human body where attention is paid to the calculations and metrics that measure a person's health and, for example, their reproductive capacities. These strategies are also implemented at the level of the population, where metrics like birth rates, death rates, epidemics, et cetera form the object of management. Biopolitics involves logics, strategies, and mechanisms for in/exclusion within the population, regulations of health and reproduction, and calculations for measuring these phenomena.

Biopolitics has contributed to an emerging research area within Romani studies that investigates how state interventions have constituted Roma as an object of population management (Marinaro, 2009; van Baar, 2017a, 2017b; Picker, 2017). Such a tactic is grounded in racialized understandings of Roma (Kóczé, 2017), and state-led interventions are part of a longer historical process of racist state control and management of Roma populations (Kóczé and Trehan, 2009). I apply these insights to gentrification in Budapest because the districts with the most well-documented cases of displacement are also the districts with the highest levels of state-owned housing (Dosa, 2009; Jelinek, 2011, 2017; Czirfusz et al., 2015; Olt and Lepeltier-Kutasi, 2018). Budapest's Eighth District has historically had both the highest proportion of state-owned housing (Czirfusz et al., 2015) and the highest population of Roma residents (Ladanyi, 1989). The connection between state-owned housing, displacement, and Roma allows for a rethinking of displacement processes in post-socialist contexts as racialized phenomena.

As Rabinow and Rose (2006) note, biopolitics is fundamentally about the politics of the human body and the population's life-sustaining functions. Drawing on Foucault's (2007) 1978–1979 lecture series and the *History of Sexuality Vol. 1* (1979), we can understand biopolitics as having two-poles: (1) 'the anatomo-politics of the body' (Rabinow and Rose, 2006: 196) and (2) the regulation of the population through its life-giving properties (Legg, 2005; Lemke, 2011). On this point, this chapter makes a slight divergence from typical biopolitical approaches by conceiving of the 'home' not only as a key node where the politics of life are made possible through the materialities of infrastructure but also as a space where positive power is enacted through state-led initiatives to build community (Foucault, 1979). While the space of the home is not biopolitics *per se*, it is inseparable from the web of biopower.

With the above in mind, I adapt the concept of biopolitics to focus on two statistical categories used as metrics in Eighth District gentrification projects: *komfortfokozat*, or 'comfort level', and Roma as a population. The empirical section of this chapter traces both of these categories through the gentrification of Budapest's Eighth District. The following section is based on three stints of fieldwork between 2016 and 2019 (a total of seventeen months), where I collected thirty interviews with activists, policy makers, and residents, as well as thousands of pages of archival data from various archives across Budapest.

Biopolitical gentrification: komfortfokozat, *the Roma population, and displacement*

Komfortfokozat data have been published in Hungary from the beginning of the twentieth century, and the metric has survived the transformation of multiple state regimes to the present. The purpose of *komfortfokozat* is to

Table 5.1 Komfortfokozat category definitions

Komfortfokozat (Comfort Level)	*Criteria*
Összkomfort (total comfort)	At least a 12 m^2 area with living space, cooking space, bathroom (with bath, or with the bath in a separate space), and toilet Electricity, plumbing, and sewage Hot water, either from inside or outside (*táv*) the building or from another central location, and supplied by an electric boiler, gas water-heater, or bath-stove Central heating system, either from inside or outside the building or from another central location or heating supply
Komfortos (comfort)	At least a 12 m^2 area with living space, cooking space, bathroom (with bath, or with the bath in a separate space), and toilet Electricity, plumbing, sewage, hot water supply Individual heating unit (gas heating, solid or oil fuelled stove heater, electric radiator)
Félkomfortos (half-comfort)	At least a 12 m^2 area with living space, cooking space, and either bathroom or a toilet Electricity, plumbing Individual heating unit
Komfort nélküli (without comfort)	At least a 12 m^2 area with living space and cooking space, and a useable toilet (or outhouse) outside the flat Some type of heating at a person's disposal The possibility to secure water
Szükség lakás (bare-necessity flat)	That which cannot be classified using the above criteria, but which has at least one room

Source: CSO (1991).

provide a standard for assessing the liveability of flats and houses according to infrastructure (see Table 5.1).

In 1991, the Central Statistics Office (CSO) published the latest data on the *komfortfokozat* for every district in the city of Budapest. The statistics showed that the Eighth District had the highest level of low-quality housing, accounting for both flats without comfort and bare-necessities flats, in the inner-city districts (see Table 5.2).

Independent studies occurring at roughly the same time also showed that the inner-city Pest Districts had a higher than average population of Roma, where the Eighth District in particular had a remarkably high proportion of Roma concentrated in the middle part of the district. In a study published over a series of articles from the late 1980s to the early 1990s, sociologist János Ladányi quantified the population of Roma by surveying teachers at 'primary schools and special schools for retarded children' (Ladányi, 1989; Ladányi, 1992: 80). Ladányi concluded that over thirty per cent of the central Eighth District is populated by Roma; that many of them live in poverty; and

Table 5.2 *Komfortfokozat* by inner-city district, 1990

District	*Total Comfort*	*Comfort*	*Half-Comfort*	*Without Comfort*	*Bare-Necessities Flats*	*Without Comfort Bare-Necessities Flats*
I.	5598	8943	626	468	239	*707*
II.	20595	17332	1320	1276	811	*2087*
V.	7953	9533	735	649	469	*1118*
VI.	2975	16532	2380	3127	558	*3685*
VII	3749	20910	3687	5463	1144	*6607*
VIII	**6487**	**18532**	**4755**	**8225**	**920**	***9145***
IX.	9982	13388	3132	5198	495	*5693*
XIII.	27286	16251	4556	7101	819	*7920*

Source: CSO (1991).

that many had come from the country-side, where access to quality educational opportunities is lower than in the city (Ladányi, 1992: 80–84). By comparison, the rest of the city had a population of roughly five per cent Roma.

The results of his study led him to connect these two data – the high population of Roma and the low-quality housing – as a crisis facing the district. In 1991, Ladányi wrote an article titled 'The Eighth District in Depth' for *Beszelő*, a magazine that published short articles from prominent politicians and scholars written for a popular audience. In this article, he highlighted the historical processes that led to the district's contemporary condition. Starting with the late nineteenth century, he recounted how the city remained on the margins and served as a home to many of the city's minorities (Ladányi, 1991). With the advent of state-socialism and the regime change that occurred after 1956, the Hungarian socialist state began to invest heavily in large housing estates in the suburbs, leaving the Eighth District badly disinvested. However, it was the arrival of the Roma population that changed the social and physical landscape:

> A new chapter in the slumification of this part of the city began with the inflow of the Gypsy population. There had always been a relatively small and localized Gypsy population living in the district. Most of them came from the upper echelons of Gypsy society [*cigányság*], and they were mostly musicians by trade. The massive influx of Gypsies occurred in the middle of the 1970s with the elimination of Gypsy and impoverished settlements, while a parallel process of 'bulldozer' urban development was occurring in the outer districts.
>
> (Ladányi, 1991: 51)

Ladányi posited that the *slumification* of the district began with the arrival of the Roma population in the 1970s. He argued that this situation had led to the outflow of 'higher status' people and an increased prevalence of criminality, placing the Eighth District on a historical trajectory that required swift action at the level of the population if it was going to be stopped. He expressed his concerns later in the article:

> If we do not manage to stop the outflow of the higher status population [*népesség*] from the district, if we cannot manage to put the brakes on the inflow of the Gypsy population, furthermore, if we continue to allow prostitution-related businesses into the district, then the possibilities for city planning will be curtailed.
>
> (Ladányi, 1991: 52)

For Ladányi, the condition of housing, measured in *komfortfokozat*, and the increasingly large Roma population created a crisis in the district that necessitated intervention.

A key intervention was to forbid large swaths of the district from privatization. This act was accomplished with the publication of the *tilalmi lista*, or blacklist, in 1992 (Józsefváros Önkormányzati Tájékoztató, 1992). With the blacklists, the Eighth District government prevented the privatization of district sections thought to be in poor condition and in need of intervention. As can be seen in Table 5.3, the primary reasons for withholding privatization rights were related to low-quality infrastructure.

Because of the blacklist, nearly 600 building in the Eighth District were prevented from privatization – the vast majority of which were within the central portion of the district, which houses the highest population of Roma (Józsefváros Önkormányzati Tájékoztató, 1992). In all, 14,000 residents were forbidden from privatizing their flats, or roughly sixteen per cent of the population of the Eighth District (CSO, 1991; Józsefváros Önkormányzati Tájékoztató, 1992). While Roma were not explicitly denied privatization,

Table 5.3 Reasons for withholding privatization

	Reason	*Number of Buildings Withheld from Privatization*
1a	City Settlement Programs – Buildings set to be demolished in accordance with the development plan	120
1b	City Settlement Programs – Slated for partial demolition of residential buildings due to the poor condition of infrastructure	36
1c	City Settlement Programs – Buildings recommended for functional change	43
2	Buildings marked as historical monuments (*műemlék*)	24
3	Previously owned by a church (*egyház*), buildings reclaimed by a church	18
4	Residential buildings with deteriorated infrastructure	360

Source: Józsefváros Önkormányzati Tájékoztató (1992).

the blacklisted flats were mostly located in the Central Eighth District, where Ladányi had shown a higher-than-average Roma population.

By the mid-1990s, a wave of evictions began in the Eighth District. In general, demolition and evictions on a case by case basis in the 1990s are difficult to pinpoint. However, interviews with three Roma activists in the 1990s indicated there were hundreds of eviction cases throughout this period (Interview 9; Interview 16; Interview 29), and one of the primary justifications for eviction was said to be unsafe building conditions (Interview 9).

During the same period, the Eighth District commissioned an organization to help spur the social and physical rehabilitation of the district called Rév8. The first major intervention planned by Rév8 was the Corvin Quarter project. As stated in early documents from the Eighth District planning commission, roughly eighty per cent of the housing stock had been owned by the district government (Rév8, 1998). The project's objective was to demolish all the publicly owned housing stock to make way for a new mixed-use residential, commercial, and consumer spaces called the Corvin Promenade. The project broke ground in 2005, on the heels of a district-led relocation project of the mostly Roma population. In all, roughly 1,100 apartments were destroyed, displacing an estimated 3,000 people (Czirfusz et al., 2015).

When asked about the condition of the Corvin Quarter prior to its demolition, most interviewees regarded the area as being in poor condition. Many long-time residents remembered the area being friendly and welcoming (Interview 20). One displaced family recalled that the primary justification for their eviction was the building's state of disrepair, despite their insistence that the building was well maintained (Interview 1). A planner in a lead position with Rév8 recalled,

> At the time, I had a big debate with a lot of sociologists who had said that we had destroyed the community. There wasn't a community there. There were a lot of new people moving into the district. In that area, it needs to be known that for many years there had been a moratorium on new construction, and it wasn't possible to do much of anything, building or otherwise, and as such the area was going downhill [*lepusztult lefelé*]. The majority of the flats were district-owned flats and were 'without comfort' [*komfort nélküli*] flats in very bad condition.
>
> (Interview 28)

Thus, the planner framed this section of the district tellingly by describing its residents as not forming a community. She also described the flats as being 'without comfort'. In this sense, she reproduced the biopolitical dualism running through much of the discourse. Stripped-bare of a 'community', the residents are rendered as a population; the flats themselves, and the 'placeness' described by earlier residents, are reproduced as statistical categories. The planner went on to say,

> The district also undertook a social programme, where young sociologists and social workers found people who were living in totally deteriorated flats. They all received proper [*rendes*] housing. Everyone who got proper housing, they all got at least comfort [*komfortos*] housing, whereas they had been living in without-comfort [*komfort nélküli*] housing previously. If they didn't have proper registration, they were given opportunities to register. No one was kicked out onto the streets. We're talking about 8,000 people. Everyone received housing – better than what they had had, better than what they would have gotten in a different type of programme. This is what the district undertook, and it was a lot of money to have this many people relocated.
>
> (Interview 28)

For this planner, the problem of low-quality housing and the lack of community could be solved with direct intervention into the built environment. In order to facilitate the project, the existing housing stock and the inherited property relationships had to be physically razed. The demolition was made possible in part because the state had maintained control over the housing stock by withholding privatization rights in the area. The demolition had the effect of removing the existing post-socialist property relations and the multiple types of tenure claims that resulted from the transition. Further still with the razing of these property relationships, the life-worlds of the inhabitants who had lived in this space were dismantled along with it. This act of material and symbolic erasure was made possible by the concentration of property in the hands of the district government.

As mentioned, the district aimed to resettle the populations that had been displaced as a result of the demolition. However, inclusion or exclusion from the resettlement programs was largely dependent on the legitimacy of the title claims on flats. One of the primary criteria for exclusion from resettlement programs was *önkényes lakásfoglalás*, translated literally as 'arbitrary flat occupation' but ultimately refers to squatting in unoccupied flats. At the time, Roma were also increasingly associated with living in squats (Kőszeghy, 2007). By the late 1990s, squatters had been stigmatized as criminals. In an interview I conducted with a district government representative who held office in the late 1990s and early 2000s, he described those living in the squats as being there 'in bad faith'. He claimed that the majority of those squatting used the flats as hubs for the sex industry and for drug dealing (Interview 7). In 1999, the Eighth District was highlighted as a hotbed for squats in the country's main centre-left newspaper *Népszbadság*. Reporting on the sentiments of a prominent right-wing politician at the time, the author wrote,

> Squatting is a dangerous act, societally speaking. The majority of those that did not possess a valid title demonstrated anti-community behaviour [*közösségellenes magatartást*] and were embittering the entire life of

> the building with their drinking and their brawling, all the while doing serious damage to the assets of the district government.
>
> (*Népszabadság*, December 14, 1999)

As a result, the district government evicted squatters and reclaimed the flats, bringing those particular properties under the control of the state.

By the end of the 1990s, Rév8 had devised a programme to not only eliminate *önkényes lakásfoglalás* but also build a community and renovate the flats in the Magdolna Quarter surrounding the Matyas Square. The project, titled the Magdolna Project, was billed as a 'social and physical rehabilitation' project. For one social worker hired for the project, the problem of *önkényes lakásfoglalás* and crime, said to be rife in the district, were intimately connected. The social worker helped implement a job-seekers programme, as well as community cooking classes, to help bring people together in the district (Interview 30). He also organized concerts with popular DJs and artists for young people in the district. The overall goal of the project was to 'rehabilitate' the neighbourhood. However, on this issue, he remarked,

> When we started to rehabilitate, like in every country, not just in Hungary but the US, too, there's a sad part and a happy part. The sad part is that people that don't have [a title] to be in the building [...]. They don't have rights to live there because they are *önkényes lakásfoglaló* [squatters], because they are just going into the buildings and just living there. Sometimes they wouldn't pay the utilities. They wouldn't pay the rent. They don't pay anything. So these people needed to leave the district.
>
> (Interview 30)

The Magdolna project exhibits a state-led, biopolitical logic of inclusion/exclusion as a means of managing the population. The district enacted a positive power campaign to foster community at the scale of the building, while simultaneously evicting squatters. These social changes coincided with a number of changes to the built environment. At Mátyás Square, the developers built a pedestrian walkway, fenced-in the green space, put a police station at the square, erected playground equipment, and landscaped the whole park to prevent free use of the space. The fence is locked up nightly at ten o'clock, and signs hang from the chain links forbidding the consumption of drugs and alcohol as well as football matches. The façades of half of the square were renovated, leaving the other half in the condition they had been in since they were built over a century ago. This square, with its half-renovated buildings and strict regulation of usage, is one of the prime examples of biopolitical gentrification in the Eighth District.

Conclusion

This chapter has taken as its central concern displacement processes in post-socialist cities, where there exists a property regime characterized as 'super-homeownership' (Stephens et al., 2016). Such a housing regime inhibits gentrification processes as they occur in Western contexts, where gentrification theory has been developed (Ghertner, 2015). With this in mind, I trace the logics of urban development in the Eighth District that began in the 1990s, in the midst of the transition from state-socialism. I highlight how experts forged a problematic conditions of flats and the higher-than-average Roma population that required intervention. I then trace these interventions through three primary actions: (1) the use of the *black list* that prevented the privatization of low-quality flats in areas with a higher-than-average Roma population; (2) the Corvin Szigony development project, which saw the displacement of over 3,000 Roma people and the destruction of 1,100 homes for the primary reason that the flats were in poor condition; and (3) the implementation of the Magdolna Project, which sought to socially and physically rehabilitate the neighbourhood through a positive power campaign of community development. In this last example, the state evicted mostly Roma residents, who were living illegally in unoccupied flats.

These processes continue into the present. In 2016 the longest-running Roma-led activist and cultural centre in the Eighth District, the Roma Parliament, was evicted from its district-owned headquarters. The primary justification from the district government was that the space had become life-threatening (*életveszélyes*), referring to the quality of the building itself (Horváth, 2016; Horváth and Zsigó, 2016). Much like the rest of the district, there had been little investment in the building since the beginning of the twentieth century, yet the Roma Parliament had operated in the building without issue for over two decades. The Roma centre's eviction, justified by the building being 'life-threatening' and unfit for living, is the latest eviction event in a longer history of a particular form of post-socialist gentrification. These gentrification projects have a dual character as they impose interventions on the home due to its poor condition, measured by the statistical category of *komfortfokozat*, with the aim of managing the Roma population. Because the home is a key node in the maintenance of the human body, and because the Roma population is taken as the object of management, I argue that displacement and gentrification in the Eighth District are fundamentally biopolitical.

References

Bernt, M. (2016). How post-socialist is gentrification? Observations in East Berlin and Saint Petersburg. *Eurasian Geography and Economics*, 57(4–5), pp. 565–587.

Blaha, M. (1992). Kilakoltatás [Eviction]. *Amaro Drom*. Trans: Jonathan McCombs.

Bodnár, J. (2004). *Fin de Millenaire Budapest: Metamorphoses of Urban Life*. Minneapolis: University of Minnesota Press.

Buckley, R. and Mathema, J. (2018). Housing privatization in Romania: An anticommons tragedy? *Economics of Transition*, 26(1), pp. 127–145.

CSO (Central Statistical Office, Hungary). (1991). 1990 Census.

Czirfusz, M., Horvath, V., Jelinek, C., Posfai, Z., and Szabo, L. (2015). Gentrification and rescaling urban governance in Budapest-Jozsefvaros. *Intersections: East European Journal for Society and Politics*, 1(4), pp. 55–77.

Derickson, K. (2015). Urban geography I: Locating urban theory in the 'urban age'. *Progress in Human Geography*, 39(5), pp. 647–657.

Dosa, M. (2009). The cultivated west, the second-class east and the Roma at the bottom of the slope: Racism and self-civilizing identity formation in contemporary Hungary. MA thesis, Central European University.

Fabula, S., Kovacs, Z., Boros, L., and Pal, V. (2017). Studentification, diversity, and social cohesion in post-socialist Budapest. *Hungarian Geographical Bulletin*, 66(2), pp. 157–173.

Ferencuhova, S. (2016). Accounts from behind the curtain: History and geography in the critical analysis of urban theory. *International Journal of Urban and Regional Research*. doi:10.1111/1468-2427.12332

Foucault, M. (1976). *History of sexuality Vol.1: The will to knowledge*. New York: Vintage Books.

Foucault, M. (2007). *Security, territory, population*. New York: Vintage Books.

Ghertner, A. (2015). Why gentrification theory sails in 'much of the world'. *City*, 19(4), pp. 552–563.

Harvey, D. (1974). Class-monopoly rent, finance capital, and the urban revolution. *Regional Studies*, 8(3–4), pp. 239–255.

Hackworth, J. and Smith, N. (2001). The changing state of gentrification. *Tijdschrift voor Economische en Sociale Geografie*, 92(4), pp. 464–477.

Horváth, C. L. (2016). Utcára tették a Roma Parlamentet [They put the Roma Parliament on the street]. *Magyar Nemzet*, 25 October.

Horváth, A. and Zsigó, J. (2016). Búcsú a Roma Parlamenttől [Farewell from the Roma Parliament]. *Népszava*, 2 November.

Hegedüs, J, Mark, K., Sarkany, C., and Tosics, I. (1996). Hungary. In: Clapham, D., Hegedüs, J., Kintrea K., and Tosics, I with Kay, H. (eds.) *Housing privatization in Eastern Europe*. London: Greenwood Press, pp. 57–77.

Heller, M. A. (1996). The tragedy of the anticommons: Property in the transition from Marx to markets. *Harvard Law Review*, 111(3), pp. 621–688.

Jelinek, C. (2011). *State-led gentrification and relocation in Budapest: Vacating a house in Ferencváros*. MA thesis, Central European University.

Jelinek, C. (2017). *Uneven development, urban policy making and brokerage: Urban rehabilitation policies in Hungary since the 1970s*. PhD thesis, Central European University.

Józsefváros Önkormányzati Tájékoztató. (1992). Házingatlanok Elidegenitésének Tilalmi Listája [Real estate privatization black list], 23 October. Trans: Jonathan McCombs.

Kóczé, A. (2017). Race, migration, and neoliberalism: Distorted notions of Romani migration in European public discourses. *Social Identities: Journal for the Study of Race, Nation, and Culture*, 24(4), pp. 459–473.

Kóczé, A. and Trehan, N. (2009). Postcolonial racism and social justice: The struggle for the soul of the Romani Civil Rights Movement in the 'New Europe'. In: Huggan, G. (ed.) *Racism, post-colonialism, Europe*. Liverpool University Press, pp. 50–77.

Kőszeghy, L. (2007). A Cigány és a Nem Cigány Lakosság Lakhatási Egyenlőtségeinek Tényezői. [Factors in unequal housing opportunities for the Gypsy and non-Gypsy population.]. In: *Disziplinák Határain Innen és Túl*. MTA Társadalomkutató Központ, pp. 11–33.

Kovács, Z., Wiessner, R., and Zischner, R. (2012). Urban renewal in the inner city of Budapest: Gentrification from a post-socialist perspective. *Urban Studies*, 50(1), pp. 22–38.

Ladányi, J. (1989). Changing patterns of residential segregation in Budapest. *International Journal of Urban and Regional Research*. doi:10.1111/j.1468-2427.1989.tb00136.x

Ladányi, J. (1991). A Józsefváros Mélyen [Józsefváros in Depth]. *Beszélő*, 3(52), pp. 50–53. Trans: Jonathan McCombs.

Ladányi, J. (1992). Gondolatok a Középső-Józsefváros rehabilitációjának társadalmi összefüggéseiről [Thoughts on the social relations of the rehabilitation of Central Eighth District]. In: Ladányi, J., Csanádi, G., Csepeli, G., Szelényi, I., and Nemeskéry, A. (eds.) *Társadalom és térszerkezeti átalakulások* [Society and spatial-structural transformations]. Originally in *Tér és Társadalom*. Trans: Jonathan McCombs.

Lees, L., Shin, H. B., Lopez-Morales, E. (2016). *Planetary gentrification*. Malden, MA and Cambridge, UK: Polity Press.

Legg, S. (2005). Foucault's population geographies: Classifications, biopolitics, and governmental spaces. *Population, Space and Place*, 11(3), pp. 137–156.

Lemke, T. (2011). *Biopolitics: An advanced introduction*. New York and London: New York University Press. Trans: Eric Frederick Trump.

Marcuse, P. (1996). Privatization and its discontents: Property rights in land and housing in the transition in Eastern Europe. In: Andrusz, G., Harloe, M., and Szelenyi, I. (eds.) *Cities after socialism: Urban and regional change and conflict in post-socialist societies*. Cambridge, MA: Blackwell Publishers, pp. 119–191.

Marinaro, I. C. (2009). Between surveillance and exile: Biopolitics and the Roma in Italy. *Bulletin of Italian Politics*, 1(2), pp. 265–287.

Nagy, E. and Timár, J. (2012). Urban restructuring in the grip of capital and politics: Gentrification in East-Central Europe. In: Csapo, T. and Balogh, A. (eds.) *Development of the settlement network in central European countries: Past, present, and future*. New York City: Springer.

Népszabadság. (1999). Össztűz a Lakásfoglalókra [Squatters Underfire]. *Népszabadság*, 14 December, p. 29. Trans: Jonathan McCombs.

OECD. (2016). HM1.3 Housing Tenures. *OECD Affordable Housing Database*. OECD Social Policy Division – Directorate of Employment, Labour and Social Affairs. http://oe.cd/ahd

Olt, G. and Lepeltier-Kutasi, L. (2018). Urban rehabilitation and residential struggles in the post-socialist city of Budapest. In: Oosterlynck, L. S., Beeckmans, L., Bassens, D., Derudder, B., Segaert, B., and Braeckmans, L. (eds.) *The city as a global political actor*. Milton Park, UK: Routledge.

Picker, G. (2017). *Racial cities: Governance and the segregation of Romani people in urban Europe*. Milton Park, UK: Routledge.

Rabinow, R. and Rose, N. (2006). Biopower today. *BioSocieties*, 1, pp. 195–217.

Rév8. (1998). Józefváros Urban Renewal Program. Rev8.

Robinson, J. (2006). *Ordinary cities: Between modernity and development*. Milton Park, UK: Routledge.

Roy, A. (2009). The 21st-century metropolis: New geographies of theory. *Regional Studies*, 43(6), pp. 819–830.

Smith, N. (1996). *The new urban frontier: Gentrification and the revanchist city*. Milton Park, UK: Routledge.

Stephens, M., Lux, M., and Sunega, P. (2016). Post-socialist housing systems in Europe: Housing welfare regimes by default? *Housing Studies*, 30(8), pp. 1210–1234.

Tuvikene, T. (2016). Strategies for comparative urbanism: Post-socialism as a deterritorialized concept. *International Journal of Urban and Regional Research*. doi:10.1111/1468-2427.12333

van Baar, H. (2011). *The European Roma: Minority representation, memory, and the limits of transnational governmentality*. PhD thesis, University of Amsterdam.

van Baar, H. (2017a). Evictability and the biopolitical bordering of Europe. *Antipode*, 49(1), pp. 212–230.

van Baar, H. (2017b). Contained mobility and the racialization of poverty in Europe: The Roma at the development-security nexus. *Social Identities: Journal for the Study of Race, Nation, and Culture*, 24(4), pp. 442–458.

Ványi, Á. (2018). Kétmilliárd forintos komplex városfejlesztés indul Józsefvárosban. [Two billion Forint Urban Development Complex Breaks Ground [indul] in Józsefváros]. *Józsefváros Havi Lap*. 18 March. https://jozsefvaros.hu/hir/5588/ketmilliard-forintos-komplex-varosfejlesztes-indul-jozsefvarosban. Trans: Jonathan McCombs

Weist, K. (2012). Comparative debates in post-socialist urban studies. *Urban Geography*, 33(6), pp. 829–849.

6 Speculating upon San Francisco's futurity

From Shell Company evictions to decolonial action

Erin McElroy

Introduction

It was on our fourth visit to Vanguard Properties headquarters in the San Francisco Mission District that we were pushed out for the last time. A couple dozen housing tenants and activists under the banners of Eviction Free San Francisco (EFSF) and the Anti-Eviction Mapping Project (AEMP) had been holding small occupations in the building at Mission Street and 21st Street, demanding the repealing of an eviction at 151 Duboce Street, home of Benito Santiago. Benito, a Filipino American housing activist, educator, dancer, and drummer had begun attending EFSF meetings in 2013, shortly after he had received an Ellis Act eviction notice by Pineapple Boy Limited Liability Company (LLC). The Ellis Act is a California state law that allows landlords to evict rent-controlled tenants for no fault of their own and then flip and sell the units for profit, speculating upon the value of a building once vacated of tenants. While those of us in EFSF were familiar with what an Ellis Act eviction was, we did not know, when Benito first showed up, who Pineapple Boy LLC was. Nor did Benito. 'I thought it was a joke!' Benito told us all at his first meeting, confused as to who his new landlord and now evictor was.

It took community-based research, linking eviction data, property ownership data, and corporate ownership data to determine that the LLC shell was really a front for Michael Harrison, co-founder of Vanguard Real Estate – the largest real estate company in the Latinx Mission neighbourhood of San Francisco. While technically this is all public data, it is not available to the public in an accessible or organized fashion, and so much of the AEMP's work involves requesting, cleaning, linking, and analysing disparate, unorganized datasets. The collective's work also involves collaboration with direct action collectives and policy reform organizations determined to thwart the onslaught of racial dispossession and eviction instigated by contemporary gentrification.[1]

The collective EFSF emerged in 2013 in the height of San Francisco's eviction epidemic correlative with the 'Tech Boom 2.0', an era beginning in 2011, following the Dot Com Boom of the late 1990s. Tech 2.0 has been in part constituted by the entangling of technocapitalism and speculative real estate in ways that amplify Dot Com Boom-era gentrification processes, leading to an array of largely racialized evictions (Mirabal, 2009; Walker, 2018). Disproportionately, evictions target tenants of colour – particularly Black and Latinx residents (McElroy, 2019; Ramírez, 2019), preying upon a deeper racial and colonial history of white wealth accumulation (Harris, 1993; Lowe, 2015; Byrd et al., 2018; Taylor, 2019). While EFSF emerged to combat evictions through direct action in the streets, at the same time, the AEMP simultaneously developed to abet direct action mobilizations with research. In this way, the AEMP has sought to utilize anti-capitalist and antiracist technological work to analyse and produce tools to counter real estate technologies of racial dispossession (Maharawal and McElroy, 2018; McElroy, 2020). As one of the AEMP's co-founders, I learned quickly that a central task of our collective would be to learn how LLC (limited liability company) shell companies work and understand how they have become so mired in speculative eviction schemas. By this, I refer to processes in which landlords speculate upon the future worth of residential units once vacated of the current occupants.

In this chapter, I draw upon AEMP analyses, EFSF participant observation, and LLC research in order to theorize the role of speculation and shell companies in San Francisco's eviction crisis. In short, I suggest that while shell companies may be used by landlords to amass financial gain and even launder money, there is also the understudied benefit of anonymity. By hiding behind obtuse-sounding LLCs names and obtuse, generally inaccessible paper trails, landlords are able to speculate and evict with impunity (Glantz, 2019), as Pineapple Boy LLC endeavoured. This speculative real estate practice builds upon a *longue durée* of racialized property violence, record keeping, and urban planning (Roy, 2017; Bhandar, 2018). Despite the amplification of racial dispossession through LLC evictions, at the same time, community-based research and direct action hold the potentiality for carving out decolonial futures.

In what follows, I begin by studying speculation as connected to broader post-2008 racialized housing financialization in the United States. I also look at how speculation is constitutive of Ellis Act evictions in California. I next offer a brief history of shell companies and LLCs, particularly as they rest upon ongoing colonial speculation. Then, drawing upon the work of the AEMP, I assess the role that LLCs play in the San Francisco housing crisis. At the same time, I note how community-based research and action can help curb their abuse. I conclude by placing real estate speculation within a deeper genealogy of settler colonialism, gesturing to the possibility of decolonizing speculation.

Speculation

Over the last several decades in the United States, finance has become less tethered to production and more enwrapped in trading money, security, and credit. While real estate has always long been reliant on finance, today it is more than ever dependent on financial markets. These can also, in turn, become contingent upon the real estate market (Aalbers, 2008: 151; Fields, 2015, 2017). This creates new forms of interdependency, in which housing becomes understood as a place to park money and invest in, rather than the space of home. The buying and selling of housing has thus become enwrapped in speculative markets, with properties understood as a means of capital accumulation, often after they are transformed from one purpose into another. In the case of San Francisco-based Ellis Act evictions, speculators buy up rental units (often in multi-unit buildings), evict the tenants, and then sell each individual unit as a tenancy-in-common – a condo-like ownership entity.

As Tsing (2011: 75–77) writes in her work on 'spectacular accumulation', often 'investors speculate on a product that may or may not exist'. Speculation is indeed spectacular, in the sense that there is an element of conjecture about it. In Bahng's (2017: 2) words, 'It is indeed a god trick to get people to mistake prophecy for truth, notional figure for value, or futurity for the future'. Might it be that speculation holds actuarial powers and that, at the same time, capitalist economics is a performativity of sorts? After all, its workings hinge upon what Marx (1909: 699) once termed 'fictitious capital', or promissory notes with no material underlying assets.

In a sense, money, like speculation, has always relied upon abstraction and speculation (Graeber, 2012). However, as the subprime mortgage crisis revealed, the loss of money through the creation of unpayable debts bears concrete histories of racial violence and dispossession (Taylor, 2019). According to Chakravartty and da Silva (2012: 8), these '"new territories" of consumption and investment have been mapped onto previous racial and colonial (imperial) discourses and practices'. As scholars of racial capitalism, notably Cedric Robinson (1983), have mapped, capitalism, from its origins, has always relied upon prior constructs of raciality. Thus, speculative real estate practices, despite their abstract logics, are constituted by, while co-constituting, race (Harris, 1993; Bhandar, 2018; Bledsoe and Wright, 2018). When it comes to the foreclosure crisis, Chakravartty and da Silva (2012: 8) write,

> Raciality, as it places the 'others of Europe' before the horizon of death, disappears with the very possibility of a relationship that would make a debt/credit situation comprehensible and hence the debt something that could be eventually paid precisely because of how it makes the colonial (African and Indigenous) other and their descendants as lacking the moral attributes (self-determination, self-transparency, and self-productivity) characteristic of persons and places (the ones they originate from) that truly embody the traits that distinguish the proper economic subject.

But what of racial dispossession in the realm of the US speculative housing market as it has impacted not only owners whose homes were foreclosed upon, but also tenants?

Speculative evictions

While foreclosure did indeed wreak havoc in the San Francisco Bay Area, particularly in Oakland and exurban spaces (Schafran, 2018; Walker, 2018; McElroy and Werth, 2019), racial dispossession in San Francisco has seen speculation operate within, but also beyond, subprime landscapes. This is not to deny that foreclosure did and indeed does decimate racialized small homeowners' lives in San Francisco (Eviction Defense Collaborative and Anti-Eviction Mapping Project, 2016: 6), but rather to acknowledge that tenant displacement in San Francisco neighbourhoods such as the Mission District are less directly tethered to 2008 global ruptures and their aftermaths (unless one wants to include Airbnb-induced displacement as a 2008 aftermath, since it, too, was birthed in San Francisco in 2008 and has since been used repeatedly as an eviction tool throughout the Mission and beyond). While the rise of Wall Street-owned single-family homes and the return to a 'rentership society' does starkly mark racialized tenancy transformations throughout the United States and California (Immergluck and Law, 2014; Fields, 2017; Levin, 2018), other speculative practices often drive tenant displacement in San Francisco.

Both 'fault' and 'no-fault' evictions have been dispossessing San Francisco residents over the last several decades, especially in the wake of the Tech Boom. While the AEMP takes a position against displacement enacted through both categories (recognizing that often tenants are evicted for the 'fault' of being poor and racialized) (McElroy and Werth, 2019), we have also found that no-fault evictions are frequently enacted by real estate speculators. Of no-fault evictions, there are three categories used most frequently: owner-move-in evictions, demolition evictions, and Ellis Act evictions. The latter was what Pineapple Boy LLC used in its attempted eviction of Benito. As the AEMP found by analysing Ellis Act evictions enacted in San Francisco between 2009 and 2013 (in partnership with the state-wide organization Tenants Together), roughly sixty per cent of 'Ellised' evictions occur within the first year of property ownership in San Francisco, and nearly eighty per cent transpire within the first five years (Tenants Together and Anti-Eviction Mapping Project, 2014: 8). This genre of eviction, in which real estate agents and landlords speculate upon the future worth of a building once vacated of rent-controlled tenants, is speculative in nature.

The Ellis Act is a 1985 California state law that permits landlords of rent-controlled buildings to exit the rental market by evicting tenants for no fault of their own. In the late 1970s, Santa Monica passed legislation to protect rent-controlled tenants from displacement, as did other California cities at the time. Shortly after, the seventeen-year-old wealthy trust-fund youth Jerome Nash inherited a rent-controlled building that happened to be filled with now rent-controlled tenants in Santa Monica. However, Nash did not want to be a landlord; he wanted to demolish the building and profit. As he elaborated in the 1984 *Nash v. City of Santa Monica*,

> There is only one thing I want to do, and that is to evict the group of ingrates inhabiting my units, tear down the building, and hold on to the land until I can sell it at a price which will not mean a ruinous loss on my investment.

Denied a demolition permit due to rent control protections, he turned to his friend in the state capital, Jim Ellis. It was Ellis who introduced the Ellis Act in 1985, so that Nash could enact his eviction.

In San Francisco, Ellis Act evictions became a common real estate technology during the Dot Com Boom as speculators targeted new tech wealth. When the bubble burst, usage petered out, only to become prevalent again during Tech 2.0. During this period, disproportionately Black and Latinx residents have been evicted, with Black tenants overrepresented by 300 per cent (Eviction Defense Collaborative and Anti-Eviction Mapping Project, 2016: 5). Post-eviction, speculators target young, white tech wealth to replace those dispossessed. As Glantz (2019) suggests, corporate landlords today target neighbourhoods of colour, so that either Black and Latinx tenants transfer their wealth to investors and their shell companies, or they are evicted altogether. Both processes abet gentrification, paving the way for white corporate wealth.

Sometimes, speculators rebrand entire neighbourhoods to incentivize this process and attract wealthier newcomers. For instance, in 2014, speculator Jennifer Rosdail rebranded the Mission as 'The Quad', home to 'Quadsters'. This toponymical rebranding, a common mapping technique amongst speculators (Rose-Redwood et al., 2019), was intended to transform the historically Latinx, working-class neighbourhood into a space for Quadsters. These, Rosdail writes, 'work very hard – mostly in high tech – and make a lot of money' (Rosdail, 2014). As she describes of this new genre of people:

> They value time greatly and want to be in a place where they can get to work quickly, meet up with their friends easily, and walk or bike instead of sitting in traffic. They take the Google Bus, the Apple Bus, or another of the reputedly less well [equipped] shuttles like the eBay Bus. They also like to eat really good food, but don't often have time to cook it. And since they work on 'campuses', and are the millennial version of the Cow Hollow 'Triangle' dwellers of the 70s and 80s, the name 'The Quad' seems a good fit.

This cartographic toponymic technology that Rosdail employed is one firmly rooted in settler geographic logics. After all, the earliest settler projects also relied upon cartographically renaming and parceling land in modes that contravened Indigenous understandings of place and space (Blomley, 2020; Tuck et al., 2014). In this way, Rosdail assembled a speculative, settler marketing sieve, 'drawing the Quad upwards as the most desirous dwelling place for Quadsters, who, according to tech hiring statistics, are 70% male and 60% white on average' (McElroy and Szeto, 2018: 20). The Quad maintains the highest rate of Ellis Act evictions to date, many of which are enacted by LLC shell companies (Anti-Eviction Mapping Project, 2018a). These shell companies allow speculators to maintain anonymity while they rip out communities of colour from their neighbourhoods and replace them with white wealth.

Shells and shellmounds

While the use of LLCs to engage in real estate speculation is a relatively new technology, real estate speculation itself is far from novel. Julian Brave NoiseCat (2018) puts it,

> Today's land rush is nothing new. For more than 200 years, there has been a run on Bay Area real estate – a relentless wave of colonization, then urbanization and now gentrification that left the Ohlone, the Bay Area's first people, landless.

Indeed, the first settler colonial projects in San Francisco were real estate projects aimed to profit and displace the Indigenous Ohlone people. Spanish colonists built twenty-one Catholic missions in California between 1776 and 1833, including in the Mission District. They stole Native lands, baptized 81,586 Indigenous 'souls', contaminated people with European disease, decimated complex ecosystems, and buried roughly 62,600 people (Madley, 2016: 36). By 1910, there were only 184 Indigenous people recorded as remaining in the Bay Area (Milliken et al., 2009: 179). While Native communities endure and thrive in the Bay Area today, they do nevertheless contend with ongoing struggles for land, ancestral bodies, and recognition. Even sacred burial sites, known as *shellmounds* (due to their incorporation of actual ocean shells) are fraught spaces. Without consent of Indigenous communities, shellmounds have frequently been slated as sites for real estate development.

Arguably, the founding of the United States was a speculative real estate project that relied upon imperial conquest, land surveying, mapping, and a series of Homestead Acts. According to Glaeser (2013: 2), 'Real estate speculation was an integral part of the 'winning of the west', the construction of our cities, and the transformation of American home life, from tenements to mini-mansions'. This speculative history is not relegated into history books; it informs settler constructs of property, race, ownership, and financialization today (Lowe, 2015; Banner, 2016; Bhandar, 2018). Indeed, in the United States, economies of colonialism and racial dispossession have become the conditions of possibility for ongoing real estate speculation. Bahng (2017: 12) suggests that even derivatives markets continue this legacy (particularly as they informed the 2008 financial crisis), thriving 'on the calculation and redistribution of projected risk and volatility ... wherein the future becomes *terra nullius*, emptied of its true uncertainty, filled with securitized risk, and sanctified by a positivist accounting of projection'. This positivist projecting informs the algorithms undergirding speculative evictions and the racial dispossession they inhere. As Byrd et al. (2018: 2) observe, 'It is precisely in this overdetermined configuration of dispossession that the speculative and crony capitalist expropriations of the crisis this time have accelerated'. Just as derivative speculations informed the racial dispossession of the foreclosure crisis, so too have shell company speculative evictions of the current moment.

The Financial Crimes Enforcement Network (FinCEN) defines shell companies as 'non-publicly traded corporations, limited liability companies (LLCs), and trusts that typically have no physical presence (other than a mailing address) and generate little to no independent economic value' (FinCEN, 2006). As was the case with Pineapple Boy LLC, most LLCs behind evictions in San Francisco are registered at seemingly random addresses, both in California and out of state. While all LLCs often allow their owners financial benefits from incorporation (for instance facilitating money laundering and tax avoidance), the anonymity that LLCs provide creates an additional incentive for speculative evictors: the ability to minimize anti-eviction dissent. In what follows here I trace LLCs' evolution into the real estate and rental markets.

While there are many real estate origin stories connected to usage of shell companies, a common reference point is 1928 Florida. It was then that Florida property owner Uly O. Thompson created the company Moline Properties as a 'security device' meant to assume outstanding mortgages. After a series of questions regarding the 'split personality' of Moline and Thompson, both became embroiled in a 1943 Supreme Court case, *Moline Properties, Inc. v. Commissioner, 319 U.S. 436*. To date, Moline Properties serves a precedent for the separate nature of properly incorporated entities and their owners (Poston and Gould, 2017).

While Moline is indeed part of the lineage of today's LLC landscape, Hamill (2005: 304) traces the genealogy of LLCs further back, to US colonial times, when corporations began vesting under state rather than federal control. Significantly, this was also the era in which Black people, stolen from their lands, became considered personal property of white settlers, used to amplify the worth of land stolen from Indigenous peoples. It was this propertied relation that Harris (1993: 1714) marks as having solidified that basis of 'whiteness as property' by the end of the seventeenth Century. During this time, slaves were considered the main form of collateral mortgage for landowners (Baptiste, 2010). In this way, land speculation and labor extraction converged, remapping white ownership and speculative ontologies.

Slave debt then incited overconfident speculation and overproduction of cotton, which was backed by banks and an emergent global finance economy (Baradaran, 2015). This speculation led to an economic crash, known as the Panic of 1837, what some consider a precursor to the subprime mortgage crisis (Roberts, 2012).

While corporations were relatively during this time, after the American Revolution (mid-eighteenth Century), they become more frequent, created through state legislature. In the mid-nineteenth century, states began issuing corporate charters for banks and transportation projects. Meanwhile, Jacksonian Democrats drafted legislation permitting sponsors the use of corporations outside of state legislative backing. This led to the passing of an array of state general incorporation laws across the country that authorized limited partnerships (Hamill, 2005: 305). Delaware established itself as a friendly place for incorporation, which today shelters numerous global corporations as well as numerous San Francisco evictors.

In 1977, during the consolidation of US neoliberalism, the first official LLC emerged through Wyoming legislature, authorizing the creation of Wyoming LLC. Led by the internationally explorative Hamilton Brothers Oil Company, the goal was to organize foreign oil business activity under US law (Hamill, 2005: 298). Prior to 1977, limited liability had come at a cost in the United States, requiring a double tax on profits. The Wyoming LLC offered a way around this, although it took the Internal Review Service (IRS) nearly eleven years to instantiate it and restrictions imposed by the Tax Reform Act of 1986 (Major, 2018: 228). After, other states began to pass similar LLC statutes, and by 1996, statutes had been passed in all fifty states (Hamill, 2005: 297). The number of LLC registrations grew throughout the 1990s, with 600,000 registered by 1999. These were involved in the finance, business service, insurance, rental and leasing, and real estate sectors (ibid.: 297). By 2010, in the aftermath of the subprime crisis, roughly twenty-five per cent of all businesses with multiple owner-investors was organized under LLCs, twice that of other forms of partnership (Major, 2018: 229). Largescale Wall Street investment firms such as Blackstone, Invitation Homes, Colony, Starwood, and Waypoint (which are now all merged) began purchasing large swaths of formerly foreclosed upon properties through their various LLCs, converting them to single family rental homes (Inglis and McElroy, 2019). By 2012, there were over 2.2 million registered LLCs in the United States (Clarke and Kopczuk, 2017: 129), many of which were involved in real estate. As I go on to explore, it takes community-based research to understand who these LLC landlords are and how they are entangled in ongoing contexts of racial dispossession and 'evictability' – an ongoing settler construct.

Community-based action and research

Aalbers (2008: 149) suggests, 'The rules and logics of Wall Street are increasingly becoming the rules and logics outside Wall Street'. While arguably the financial logics of Wall Street and big corporations alike have been studied and applied by real estate speculators to amass financial gain and launder money, here I suggest that there is another benefit at play in LLC incorporation and property speculation: stymying tenant organizing and eviction research. In an article for *The Business Times*, Badger (2018) writes,

> L.L.C.s shield property owners from personal liability while obscuring their identities. In some cases, so much anonymity also enables money laundering, and it can mean that tenants struggle to hold landlords accountable, that cities fail to fix blight and that researchers can't answer basic questions about the housing market.

She also observes, 'For renters, or tenants mired in rent-to-own contracts, these entities mean they often don't know whom they're dealing with – or who's evicting them'. For instance, like many San Francisco evictees, Benito had been paying his rent on time for years to an aging landlord. One day, he discovered that the landlord had sold the building and that, going forward, his check should be paid to the new owner, Pineapple Boy LLC. He recounts in an interview conducted with Manissa Maharawal as part of the AEMP's oral history project, *Narratives of Displacement and Resistance* (Maharawal and McElroy, 2018),

> The next thing I know I get a letter from [my former landlord]. She says, 'Oh we sold the place. We sold it to Pineapple Boy LLC, this is where you'll send your checks, don't send them to me anymore.' That was November. Then I get from Pineapple Boy LLC, a letter from them, introducing themselves, and requesting us to – me and cotenants – to buy us out for $20,000 to move out within 30 days, something like that. … So, I'm talking with my co-tenants here. The gentleman works for the city and county of San Francisco. He says, 'Benito the housing market … is $4,000 a month for one bedroom. So, this $20,000 is not going to last a year'.

While shedding light on the dystopic nature of the San Francisco rental market, where on average units do cost between USD 3,000 and USD 4,000 per month, Benito's cotenant also suggested that they fight the eviction and not take the buyout. As a senior with disabilities and fixed income, this offered hope, as, 'Really, there was nowhere to go'. And so, Benito found his way to EFSF. But how does one fight a shell company evictor?

In conducting this research around Pineapple Boy LLC, the AEMP discovered that its owner, Michael Harrison, had already Ellis Act evicted two other buildings (nine units in total) in San Francisco. We also discovered that he was a member of the National Association of Realtors, the California Association of Realtors, the San Francisco Association of Realtors, the Marin Board of Realtors, and Tahoe Sierra Board of Realtors. With this information, EFSF then planned a series of protests, beginning at Vanguard. On our fourth visit, we were pushed out so forcefully by company employees that Vanguard feared legal retaliation and promptly rescinded the eviction. Thus, through a combination of community-based research and direct action, Benito and his cotenants were able to remain housed in what is now considered the most expensive and cutthroat rental market in the United States (Walker, 2018).

While Benito, with the mutual aid of EFSF's direct action and AEMP research, was able to fight Pineapple Boy LLC and remain housed, many tenants who receive LLC-led eviction notices from LLCs do not have the connections to do so, never mind the resources to determine who their evictor actually is. While there is little national data on LLC ownership due to their very nature, a 2015 Rental Housing Finance Survey conducted by the Census Bureau and the Department of Housing and Urban Development found that roughly fifteen per cent of all US properties and one-third of all rental units are owned by LLCs, limited partnerships, or limited liability partnerships (United States Census, 2018). In 1991, ninety-two per cent of all US rental units were owned by individuals whose names could be easily found in public records (Bureau of the Census, 1996), but by 2015, that percentage had shrunk to seventy-four per cent due to LLCs. These numbers in part prompted the US Treasury Department in 2016 to order title insurance firms to report the real names of those behind luxury residential purchases in New York City, Miami, Broward, Palm Beach, Los Angeles, San Francisco, San Diego, and San Antonio (Solomont, 2018). The idea was to crack down on real estate money laundering. However, while FinCEN and government bodies have access to newly collected information, it remains unavailable to the public. In 2018, a bill was passed in California to mandate the investors who own more than hundred properties register their investment activity with the state. However, there remains much opposition from the California Apartment Association, and the data is not yet available to the public (California Legislative Information, 2018). Thus, LLC activity continues, as evictions accumulate.

As the AEMP has found, of the 261,911 registered landlords in San Francisco, 132,625 own more than one property (Anti-Eviction Mapping Project, 2018b). The AEMP has also created a tool for users to look up addresses and determine prior eviction history, to pledge not to rent or buy from said speculators (Anti-Eviction Mapping Project, 2018d). It is also in the midst of developing another tool in partnership with Portland's Mapping Action Collective, *EvictorBook*, which links eviction data, corporate entity data, and property parcel information. The AEMP's New York City chapter also has created an interactive tool in collaboration with JustFix.nyc that maps the city's worst evictors, also noting the banks funding them (JustFix.nyc and Anti-Eviction Mapping Project, 2019). Further, the AEMP's website contains dozens of pages for individual speculators, with information on who they are, related evictions, and their various LLCs. Data and tools such as these are just some of many means of fighting for a futurity beyond that of speculative evictions. Meanwhile, JustFix.nyc has created a tool similar to EvictorBook, *Who Owns What* (2020), and has helped SAJE in Los Angeles develop a similar *Own It* tool (2020). In Detroit, Property Praxis has created a tool to link LLC to ownership data, available to the public (Akers and Seymour, 2018). Projects such as these create data to empower anti-displacement struggles and policy reform. This work is difficult, and it involves obtaining, cleaning, and connecting datasets purposefully disaggregated by city and corporate data bodies.

As AEMP research has found, Pineapple Boy LLC, while perhaps one of the more absurdly titled LLCs, is actually quite small compared to other speculative evictors. For instance, Veritas, San Francisco's largest landlord, owns at least 5,802 units. Founded by Yat-Pang Au, it operates over hundred LLCs financed by Wall Street banks such as Goldman Sachs. Two of them, Russian Hill Associates LLC and 108 Albion LLC, have Ellised three buildings, and buy-outs and fault evictions have taken place in a number of others (Anti-Eviction Mapping Project, 2019). Numerous tenants have experienced harassment as well. Despite this, they have been successfully organizing resistance, supported by the tenant justice counselling organization, the Housing Rights Committee.

Meanwhile, the large investment firm Urban Green has Ellised dozens of units through its LLCs. Urban Green is a subsidiary of a larger company based out of Colorado, Cornerstone Holdings. The family behind both companies, the McCloskeys, is tied to old political money, a number of conservative projects, politicians, and think tanks. Infamously, in 2013, Urban Green's 55 Dolores Street LLC Ellised the tenants at 55 Dolores Street, shortly after purchasing the building. One of the tenants was then ninety-seven-year-old Mary Elizabeth Phillips, who continued to fight the eviction until the age of 100, when she passed away. Her fight was empowered by widespread community organizing against Urban Green. Despite these efforts, 55 Dolores Street has now been flipped and sold as tenancy-in-common condos, and a new Whole Foods has been constructed across the street. As AEMP research has found, all of the new condo owners are in the tech industry (Anti-Eviction Mapping Project, 2018c).

Decolonizing speculation

Pineapple Boy LLC engaged in the act of colonizing the future the moment it was conceived. Poignantly, pineapples themselves are symbols of colonial hospitality and can be traced to the Spanish colonization of the Carib. They were also part of slave plantation economics in the American South, when they became 'antebellum "bling" status symbols' used to illuminate social hierarchy (Dawson and Mitchell, 2017: 120). In the Tech 2.0 era, this bling status has become the LLC itself, which can be mobilized to materialize an updated version of land speculation, albeit with new forms of strategic anonymity. As I have shown throughout this chapter, LLCs comprise a modern speculative technology aimed at sheltering real estate abuse in the form of ongoing settler expropriation. Speculators make their real estate computations not only by speculating upon post-eviction housing value, but also by calculating that their LLC-enabled incognito will permit impunity.

Although the future is an already occupied space, it is never completely foreclosed upon. Benito's success in collaboratively fighting his eviction makes this clear. Other futures are possible outside of capitalist and colonial speculations. These speak to what Bahng (2017: 9) describes as 'decolonizing the future in financial times'. In other words, despite the accumulation of speculative real estate calculi and racialized evictions, other futures are also being conjured and materialized.

After all, the future that Benito, EFSF, and the AEMP together speculated upon – one in which Pineapple Boy LLC would not displace 151 Duboce Street – is one of many anti-capitalist, antiracist, and decolonial futures being conjured in the gentrifying Bay Area. For instance, in 2012, Corrina Gould (whose ancestors are among the 184 Bay Area Indigenous people who survived into 1910) established the Sogorea Te Land Trust with her collaborator, Johnella LaRose. The land trust's aim is to reclaim Indigenous Ohlone Land. Its Shuumi Land Tax was established for non-Indigenous residents to pay the land's First People reparations. Unlike speculative LLC evictors, who bypass taxation and who endeavor to maintain anonymity, the Shuumi Land Tax begets settlers to acknowledge the violence of settler speculation and to decolonize its futurity. Actions such as Benito's, along with projects such as Gould's and LaRose's, collectively speculate upon futures beyond those of speculative dispossession, ones of housing and racial justice.

Note

1 This is not to suggest that all evictions in the San Francisco Bay Area are a result of contemporary gentrification. On the contrary, since the first moment of colonial violence until the present, there is a deep palimpsest of racial dispossession that lays the ground for the present (Noisecat, 2018; Ramírez, 2019; Roy, 2017). Further, it is important to acknowledge that historic and contemporary forms of racial dispossession do not impact all Bay Area geographies evenly (McElroy and Werth, 2019).

References

Aalbers, M. (2008). The financialization of home and the mortgage market crisis. *Competition and Change*, 12(2), pp. 148–166.

Akers, J., and Seymour, E. (2018). Instrumental exploitation: Predatory property relations at city's end. *Geoforum*, 91, pp. 127–140.

Anti-Eviction Mapping Project. (2018a). *All San Francisco evictions, 1997–2018.* [Online]. Available at: https://antievictionmap.com/#/sf-evictions/ [Accessed 12 May 2019].

Anti-Eviction Mapping Project. (2018b). *Corporate owners.* [Online]. Available at: https://corporate-owners-sf.netlify.com [Accessed 12 May 2019].

Anti-Eviction Mapping Project. (2018c). *House of pain / Casa Dolores.* [Online]. Available at: https://antievictionmap.com/after-the-ellis-eviction [Accessed 12 May 2019].

Anti-Eviction Mapping Project. (2018d). *Pledge map.* [Online]. Available at: http://www.antievictionmappingproject.net/pledge [Accessed 12 May 2019].

Anti-Eviction Mapping Project. (2018e). *Santa Monica Ellis Act evictions.* [Online]. Available at: https://www.antievictionmappingproject.net/santamonica.html [Accessed 12 May 2019].

Anti-Eviction Mapping Project. (2019). *Veritas.* [Online]. Available at: https://www.antievictionmap.com/veritas [Accessed 20 September 2019].

Badger, E. (2018). Anonymous owner, L.L.C.: Why it has become so easy to hide in the housing market. *New York Times*, 30 April 30. [Online]. Available at: https://www.nytimes.com/2018/04/30/upshot/anonymous-owner-llc-why-it-has-become-so-easy-to-hide-in-the-housing-market.html [Accessed 12 May 2019].

Bahng, A. (2017). *Migrant futures: Decolonizing speculation in financial times.* Durham, NC: Duke University Press.

Banner, S. (2016). *Speculation: A history of the fine line between gambling and investing.* Cambridge: Oxford University Press.

Baptist, E. (2010). Toxic debt, liar loans, and securitized human beings: The panic of 1837 and the fate of slavery. *Common-Place: The Interactive Journal of Early American History*, 10(3), pp. 651–677.

Baradaran, M. (2015). *How the other half banks: Exclusion, exploitation, and the threat to democracy.* Cambridge: Harvard University Press.

Bhandar, B. (2018). *Colonial lives of property: Law, land, and racial regimes of ownership.* Durham, NC: Duke University Press.

Bledsoe, A. and Wright, W. J. (2019). The anti-Blackness of global capital. *Environment and Planning D: Society and Space*, 37(1), pp. 8–26.

Blomley, N. (2020). Precarious territory: Property law, housing, and the socio-spatial order. *Antipode*, 52(1), pp. 36–57.

Bureau of the Census. (1996). *Who owns the nation's rental properties*? [Online]. Available at: https://www.census.gov/prod/1/statbrief/sb96_01.pdf [Accessed 12 May 2019].

Byrd, J. A., Goldstein, A., Melamed, J., and Reddy, C. (2018). Predatory value: Economies of dispossession and disturbed relationalities. *Social Text*, 36(2 (135)), pp. 1–18.

California Legislative Information. (2018). *AB-354 Institutional investors: Housing.* Available at: http://leginfo.legislature.ca.gov/faces/billNavClient.xhtml?bill_id=201720180AB354 [Accessed 12 May 2019].

Chakravartty, P. and Da Silva, D. F. (2012). Accumulation, dispossession, and debt: The racial logic of global capitalism – An introduction. *American Quarterly*, 64(3), pp. 361–385.

Clarke, C. and Kopczuk, W. (2017). Business income and business taxation in the United States since the 1950s. *Tax Policy and the Economy*, 31(1), pp. 121–159.

Dawson, J. and Mitchell, J. T. (2017). Bitter sweets: Mapping pineapples, hospitality, and slavery. *The Geography Teacher*, 14(3), pp. 118–129.

Eviction Defense Collaborative and Anti-Eviction Mapping Project. (2016). *Eviction report, 2015.* [Online]. Available at: https://static1.squarespace.com/static/52b7d7a6e4b0b3e376ac8ea2/t/5b1276fc8a922da05029225d/1527936777268/EDC_2015.pdf [Accessed 12 May 2019].

Fields, D. (2015). Urban struggles with financialization. *Geography Compass*, 11(11), e12334.

Fields, D. (2017). Unwilling subjects of financialization. *International Journal of Urban and Regional Research*, 41(4), pp. 588–603.

FinCEN. (2006). *Statement of associate director, for regulatory policy and programs*, November 14. [Online]. Available at: https://www.fincen.gov/news/testimony/statement-associate-director-regulatory-policy-and-programs-jamal-el-hindi-financial [Accessed 12 May 2019].

Glaeser, E. (2013). A nation of gamblers: Real estate speculation and American history. *American Economic Review*, 103(3), pp. 1–42.

Glantz, A. (2019). *Homewreckers: How a gang of wall street kingpins, hedge fund magnates, crooked banks, and vulture capitalists suckered millions out of their homes and demolished the American dream.* New York: Custom House.

Graeber, D. (2012). *Debt: The first 5000 years.* London: Penguin.

Hamill, S. P. (2005). The story of LLCs: Combining the best features of a flawed business tax structure. In: Bank, S. and Stark, L. (eds.) *Business tax stories.* New York: Foundation Press, pp. 295–315.

Harris, C. I. (1993). Whiteness as property. *Harvard Law Review*, 106(8), pp. 1707–1791.

Immergluck, D. and Law, J. (2014). Speculating in crisis: The intrametropolitan geography of investing in foreclosed homes in Atlanta. *Urban Geography*, 35(1), pp. 1–24.

Inglis, A. and McElroy, E. (2019). California Wall Street landlords in 2019. *Tenants Together*, November. [Online.] http://www.tenantstogether.org/sites/tenantstogether.org/files/California%20Wall%20St%20Landlords%20in%202019.pdf [Accessed 13 May 2020].

JustFix.nyc. (2020). *Who Owns What.* [Online.] https://whoownswhat.justfix.nyc/en/ [Accessed 13 April 2020].

JustFix.nyc and the Anti-Eviction Mapping Project. (2019). *Worst Evictors NYC.* [Online.] Available at: https://www.worstevictorsnyc.org [Accessed 13 April 2020].

Levin, M. (2018). Data dig: Big investment firms have stopped gobbling up California homes. *CalMatters*, 4 April. [Online]. Available at: https://calmatters.org/articles/data-dig-big-investment-firms-have-stopped-gobbling-up-california-homes [Accessed 12 May 2019].

Lowe, L. (2015). *The intimacies of four continents.* Durham, NC: Duke University Press.

Madley, B. (2016). *An American genocide: The United States and the California Indian catastrophe, 1846–1873.* New Haven, CT: Yale University Press.

Maharawal, M. and McElroy, E. (2018). The Anti-Eviction Mapping Project: Counter mapping and oral history toward Bay Area Housing justice. *Annals of the American Association of Geographers*, 108(2), pp. 380–389.

Major, A. (2018). The new capitalist rich: Corporate organizational form and the political economy of US income inequality. *Critical Historical Studies*, 5(2), pp. 209–236.

Marx, K. (1909). *Capital, Vol. III.* Chicago, IL: Charles H. Kerr.

McElroy, E. (2019). Data, dispossession, and Facebook: Techno-imperialism and toponymy in gentrifying San Francisco. *Urban Geography*, 40(6), pp. 826–845.

McElroy, E. (2020). Property as technology. *City*, 24(1–2), pp. 112–129.

McElroy, E. and Szeto, A. (2017). The racial contours of YIMBY/NIMBY Bay Area gentrification. *Berkeley Planning Journal*, 29(1), pp. 7–44.

McElroy, E. and Werth, A. (2019). Deracinated dispossessions: On the foreclosures of 'gentrification' in Oakland, CA. *Antipode*, 51(30), pp. 878–898.

Milliken, R., Shoup, L., and Ortiz, B. (2009). *Ohlone/Costanoan Indians of the San Francisco Peninsula and their neighbors, yesterday and today*. Oakland, CA: Archaeological and Historical Consultants.

Mirabal, N. R. (2009). Geographies of displacement: Latina/os, oral history, and the politics of gentrification in San Francisco's Mission District. *The Public Historian*, 31(2), pp. 7–31.

Noisecat, J. B. (2018). 'It's about taking back what's ours': Native women reclaim land, plot by plot. *Huffington Post*, 22 March. [Online]. Available at: https://www.huffingtonpost.com/entry/native-women-oakland-land_us_5ab0f175e4b0e862383b503c [Accessed 2 May 2019].

Poston, B. and Gould, M. (2017*)*. Applicability of Moline properties to S corporations. *Memorandum*, October 24. [Online]. Available at: https://www.irs.gov/pub/irs-wd/201747006.pdf [Accessed 2 May 2019].

Ramírez, M. M. (2019). City as borderland: Gentrification and the policing of Black and Latinx geographies in Oakland. *Environment and Planning D: Society and Space*, https://doi.org/10.1177/0263775819843924

Roberts, A. (2012). *America's first great depression: Economic crisis and political disorder after the Panic of 1837*. Ithaca, NY: Cornell University Press.

Robinson, C. J. (1983). *Black Marxism: The making of the Black radical tradition*. Chapel Hill: University of North Carolina Press.

Rosdail, J. (2014). *The Quad*. [Online]. Available at: https://www.jenniferrosdail.com/tag/the-quad [Accessed 2 May 2019].

Rose-Redwood, R., Vuolteenaho, J., Young, C., and Light, D. (2019). Naming rights, place branding, and the tumultuous cultural landscapes of neoliberal urbanism. *Urban Geography*, 40(6), pp. 747–761.

Roy, A. (2017). Dis/possessive collectivism: Property and personhood at city's end. *Geoforum*, 80, pp. A1–A11.

SAJE. (2020). *Own It*. [Online]. https://www.ownit.la/ [Accessed 12 May 2020].

Schafran, A. (2018). *The Road to resegregation: Northern California and the failure of politics*. Berkeley: University of California Press.

Solomont, E. B. (2018). Treasury issues new LLC disclosure law. *The Real Deal*, 17 May. [Online]. Available at: https://therealdeal.com/2018/05/17/treasury-issues-new-llc-disclosure-law [Accessed 2 May 2019].

Taylor, K. (2019). *Race for profit: How banks and the real estate industry undermined Black homeownership*. Chapel Hill: University of North Carolina Press.

Tenants Together and the Anti-Eviction Mapping Project. (2014). *The speculator loophole: Ellis Act evictions in San Francisco*. [Online]. Available at: https://static1.squarespace.com/static/52b7d7a6e4b0b3e376ac8ea2/t/5b127473f950b792dfdb3c97/1527936162241/Speculator+Loophole+2014.pdf [Accessed 2 May 2019].

Tuck, E., McKenzie, M. and McCoy, K. (2014). Land education: Indigenous, post-colonial, and decolonizing perspectives on place and environmental education research. *Environmental Education Research*, 20(1), pp. 1–23.

Tsing, A. L. (2011). *Friction: An ethnography of global connection*. Princeton, NJ: Princeton University Press.

United States Census. (2018). *Rental housing finance survey*. [Online]. Available at: https://www.census.gov/data-tools/demo/rhfs/rhfs.html?s_appName=rhfs&s_tableName=TABLE2&s_type=2&s_byGroup1=6 [Accessed 2 May 2019].

Walker, R. (2018). *Pictures of a gone city: Tech and the dark side of prosperity in the San Francisco Bay Area*. Oakland, CA: PM Press.

7 Keeping out the poor

Banishment as an urban renewal strategy

Guy Baeten and Carina Listerborn

Introduction

Over the past decades, we have witnessed increasingly repressive policies towards all sorts of urban outcasts, alongside ongoing efforts to attract population segments deemed more favourable. While at some point attempts to keep the poor away from city centres were confined to the poorest and most visible segments, such as the homeless and rough sleepers (see e.g., Mitchell, 1997), removal efforts have gradually expanded and now include much larger groups of unwanted people. In the wake of New Orleans's urban renewal policies after Hurricane Catrina, Lees et al. (2008) discerned a 'fourth wave of gentrification' wherein systematic state-led social cleansing has come to play a fundamental role in gentrification processes. The 'literal and figurative effacing of the proletariat in the city' (Wacquant, 2009: 199) can take many forms – varying from legal bans from parks and other places (Beckett and Herbert, 2009, 2010, 2017), systematic racialized evictions (Roy, 2017), or one-way tickets for homeless families (Bosman and Barbaro, 2009). Since 2006, the so-called Rotterdam Wet (van Gent et al., 2017) allows Dutch cities to refuse a residence permit in certain neighbourhoods for people who have lived in the metropolitan region for less than six years and who have no income from work, pensions, or student loans. In Denmark, the Ghetto Plan seeks to stop municipalities from moving people on benefits into public housing in areas officially designated as 'ghettos' (Simonsen, 2016; Milne, 2018).

In recent years, the city of Landskrona, a small town in the south of Sweden, has implemented a housing policy that is mainly aimed at preventing poor people from migrating to (or within) the city. The pressure on Landskrona's housing market is low compared to nearby larger cities and university towns, where housing prices and rent levels result in an increasingly excluding housing market. In the absence of these 'market forces', Landskrona's city authorities have embarked on a set of excluding policy measures. We argue that these measures herald a new era of urban population management and urban renewal in Sweden. These measures also imply that our conventional conceptual apparatus to capture social change in the city – in particular, the management of poor population segments – does not lend itself very well to understand contemporary population and renewal policies in the city of Landskrona. We are not witnessing processes of gentrification, there are no legal bans addressing poor people, there is no outright displacement, and poor people are not evicted from their homes. The new policies are not even *directly* aimed at restricting the freedom of movement of poor population segments like in the examples above; rather, they are aimed at restricting the freedom of private landlords and, in that way, indirectly overseeing the movement of poor people. We find the term 'banishment' (Roy, 2017) particularly useful to grasp the new urban renewal tactics that are unfolding in Landskrona.

Based on an empirical study of Landskrona's rental policies, this chapter aims to develop *banishment* as a concept to capture a certain variation of displacement that is currently difficult to place in the available conceptual apparatus.

City-wide rental restrictions as banishment

In the aftermath of the shipyard closure in 1980, the city of Landskrona initially tried to compensate for the inner-city White flight and resulting housing vacancy by welcoming, assisting, and housing newly arrived refugees from Lebanon, Eastern Europe, and the Balkan countries. This changed the social demography of the city significantly, triggered a range of issues regarding income and education levels, and racialized crime and safety perceptions. These problems were originally met with a series of *social* programmes and a range of 'soft' measures to help people get into employment or improve educational achievement (Baeten and Listerborn, 2015). The subsequent electoral successes of the extreme right-wing party, Sweden Democrats, together with growing discontent amongst shopkeepers and business leaders alike, triggered an entirely new approach to the city's problems.

The game-changing Crossroads Plan of 2012 (Landskrona Stad., 2012) no longer prioritized social answers to social problems. Instead, it considered high levels of affordable rental housing – for decades the cornerstone of social-democratic housing policies and the cornerstone of the nationalist socialist *folkhem* ideology – as the single root cause of all problems. The problem of poverty was now redefined as a problem of poor peoples' presence in the city, enabled by specific housing market features. Changing those features would ban poor people from the city. In order to achieve this goal, the city authorities successfully disseminated a storyline wherein 'unserious landlords' rent out to 'benefit-dependent' tenants, a process that is regarded to cause crime, low education levels, unemployment, benefit dependency, unsafety, and a weak tax base.

Through its subsidiary Landskrona Stadsutveckling AB, a public-private partnership of municipal authorities and (major) private landlords, the city not only issued a warning for aspiring immigrants on social benefits or low income, but it also issued a city-wide rental policy in 2013 (renewed in 2015) that would apply to both private and public landlords alike. This policy prevents landlords from renting out to persons on social benefits and to persons who would spend more than one-third of their income on rent (alongside some other restrictions). These restrictions proved to have potentially far-reaching effects. First, the policy deters persons on social benefits or low income from moving to Landskrona from nearby municipalities. The policy was accompanied by verbal violence from city officials and politicians. In a local newspaper interview in 2012, the Mayor declared 'My message to all welfare benefit recipients is, do not move to Landskrona. If you have a problem, then please ask elsewhere where you are more likely to get attention' (Lönnaeus, 2012). In the same interview, he stated that 'we have a city centre characterized by social benefit dependency'. One of the architects of the Crossroads Plan stated in a newspaper interview in 2011 that the city has allowed in 'a category of people we do not want' and that these unserious tenants should be 'returned to the municipalities where they come from' (Brant, 2011).

Second, the policy deters landlords from renting out to poor population segments. Twenty-three landlords have now officially subscribed to the city's rental policy, covering seventy-five per cent of the total rental housing stock (Landskrona Stadsutveckling, 2017), leading to a reduction in both the volume of housing stock available to the poor and the number of 'unserious landlords'. The policy is supported by real and verbal intimidation. The CEO of Landskrona Stadsutveckling AB declared in an interview,

> There are some property owners who are not interested. We try to identify those who do not care, and we will mobilize the authorities [such as environmental inspection and fire safety control]. If you only want a property as a cash-cow, then that should cost; that should not be fun. We will use all means possible to force property owners to jump on the development bandwagon.
>
> (Petterson, 2013)

In other words, all landlords who do not comply with the rental policy are deemed 'unserious', and the city will use its powers to guard fire safety and environmental standards to threaten 'unserious' landlords with potential compulsory purchase orders if they do not respect the city's rental policy restrictions. 'Unserious' is of course a very vague category and therefore renders significant power to the city authorities, who can both decide which landlord is not serious *and* take the necessary measures accordingly. Thus, the city designs the policy, acts as a judge, and implements repressive measures if needed, all through its newly established public-private company Stadsutveckling AB. The public-private nature of the implementing body keeps these actions outside the 'normal' democratic procedures of accountability to and control by the peoples' representatives in the City Council.

Limited effects

Five years after the policy was implemented, the city tax base has not improved. If the Swedish average tax base equals 100, then Landskrona's tax base was 84 in 2011 and 83 in 2018 (SCB, 2019). The amount of persons (aged 18–64 years) on social benefits has decreased by 14.5 per cent between 2010 and 2017, but so has the national figure (8.8 per cent) (Socialstyrelsen, 2018). The 2017 Annual Report of Landskrona Stadsutveckling announced that the total amount of social benefit expenditures has decreased by twenty-seven per cent between 2013 (the beginning of the new policy) and 2017, in contrast with the upward trend between 2007 and 2012. This insinuates that the policy has thus far been effective. However, decreasing expenditures reveal nothing about the amount of poor people in the city or the number of people on social benefits: it can simply be the result of austerity policies or stricter admission policies. In fact, the share of inhabitants on low income – the main group potentially in need of social benefits – has remained stable over the past few years (from 14.3 per cent in 2014 to 14.4 per cent in 2016) (Socialstyrelsen, 2018). Statistical evidence points at minor migratory movements of benefit-dependent people: Before the rental policy, ten to fifteen people on social benefits would move into Landskrona per month; after the policy, zero to six would move (Petterson, 2013). Thus, the policy has had some 'success', but the sheer numbers of 'problem immigrants' (before and after the policy implementation) renders the 'benefit-dependent' as a major source of Landskrona's problems a misleading myth. A recent upswing in organized crime (between December 2018 and August 2019, eight bombs detonated in the inner city) has dented the fragile image of restored peacefulness in the inner city.

It is clear that the attempts to change the housing market structure through targeting 'unserious landlords' while keeping the 'benefit-dependent' at bay have thus far had very little aggregate effect. Nevertheless, they have had drastic consequences for all those who were left with no place to go after the rental restrictions had expelled them effectively from reasonable access to housing, or those who are now under pressure from insecurity, anxiety, and temporality.[1] There is evidence that inhabitants who cannot live up to the income restriction are told by city officials to leave the city of Landskrona, which can be regarded as de facto banishment (Whrethov, 2013a,b). We have argued elsewhere (Baeten and Listerborn, 2015; Listerborn and Baeten, 2016) that the reluctance of municipalities to accept citizens on low income may lead to the emergence of 'city-less citizens' who have nowhere left to go since no municipality is prepared to enrol them. Recently, the National Board of Housing (2014) has lauded the Landskrona model as an example for other municipalities. Perhaps it is not the current (lack of) results of Landskrona's housing policy but its potential to institutionalize and systematize housing precariousness through state policies that primarily deserves the attention of urban scholars.

While ultimately aimed at keeping poor population segments at bay, the city-wide rental policy does not install measures directly targeting that population. There is no straightforward displacement and no evictions or legal restriction bans imposed on the poor. Neither are there immediate signs of gentrification hinting at the replacement of poorer population segments with more wealthy ones as we have witnessed in many other places over the past decades. We would like to propose this as a new type of displacement that clearly resonates with Marcuses's (1985) concepts of exclusionary displacement and displacement pressure but that is perhaps best captured by the term (racial) 'banishment' (Roy, 2017). We would like to carefully put forward the claim that the Landskrona case is but one of several similar examples of contemporary urban development policies that do not explicitly aim to attract wealth but, rather, to banish want. This would entail a shift from a focus on 'positive' urban renewal policies (attracting the well-off) towards a focus on 'negative' urban renewal policies: removing and keeping low-income groups at a distance – namely, banishment.

Banishment as practice

Banishment is a practice that goes back to ancient times (Armstrong, 1963) and has, over the centuries, taken on many shapes and forms across the globe. Banishment tactics, legally or informally, have typically been used to deal with persons deemed criminal, as with the deportation of petty criminals to North America, and later Australia, in the days of the British Empire. Throughout history, banishment has most commonly been deployed to remove the poor, in all their manifestations, from (inner) cities. In fact, most 'criminals' banished to the colonies under British rule were primarily starving urban citizens forced into theft (Coldham, 1992; Vaver, 2011). Struggles over urban presence in and occupation of inner-city properties and public spaces may always have been there, but according to Blomley (2004), they have reached a new intensity in contemporary times as cities strive to attract tourists, wealthier inhabitants, and inward investment: 'Development-driven displacements ... are sustained by an increasingly "revanchist" approach to marginalized urban populations, as well as "common-sense" assumptions about property, such as the "naturalness" of displacement, or the beneficial effects of middle-class ownership in poor neighbourhoods' (Blomley, 2004: xviii). Gutsche (2015) argues, the contemporary obsession with urban boosterism is more than just the promotion of 'good news' about certain places in order to attract 'the right people'. It equally holds 'restrictive and divisive purposes, which in these cases allowed "positive news" to perform its own type of banishment by restricting particular voices and perspectives from socio-spatial rhetoric'. Urban boosterism always implies the telling of shadow stories, namely those of spatial stigma (Wacquant, 2009) and regulation that restricts particular social groups from participating in inner-city spaces (be they public or private). Every boosterist story about what is wanted in the inner-city by definition implies a complementary (untold) story of what is *not* wanted. Banishment, then, is the flipside of boosterism, and socio-demographic change in the inner-city is the result of the dialectical interplay of boosterism and banishment, or of stories about the wanted (to be attracted) and the unwanted (to be banished). Gutsche (2015) goes as far as to speak of boosterism *as banishment.*

Banishment, like more conventional concepts such as 'displacement' and 'eviction', sends out a clear message about the unwantedness of individuals or groups in certain places. Unlike actual displacement and eviction, it is not only a physical act of removal but it can also be mobilized as a threatening *discursive* instrument that may, or may not, have material consequences, depending on the reactions of the unwanted population to the ban. It highlights the 'displaceability' of certain urban populations, whether actually displaced or evicted or not. Banishment underscores the ability of the state to displace, regardless of implementation. Displacement means that 'forces outside the household make living there impossible, hazardous, or unaffordable' (Hartman et al., 1982: 3), while eviction is primarily an act of legal enforcement. Banishment does not possess the same forcefulness or enforceability. Those who are banished are supposed to internalize the notion that they are no longer welcome in certain places, and they will have to comply with it, but they possess a certain freedom not to do so, or not yet. For example, research in Seattle (Beckett and Herbert, 2009, 2010, 2017) shows how people subjected to a legal ban from parks or drug dealing places often ignore the ban at a certain risk because they need to be in those places to have access to food, precarious employment, social networks, and basic services.

Banishment may not have the same forceful effect, but the very pressure of the ban of course has lasting consequences for the banned, in pretty similar ways to what Marcuse (1985) has called 'displacement pressure'. People who are banned but still in place do not suffer from actual displacement but from the anxieties, uncertainties, insecurities, and temporalities that arise from possible displacement in the near future. Banishment can also be regarded as a variant of Marcuse's (1985) 'exclusionary displacement' or the impossibility of certain groups to move into a place or a neighbourhood due to rent levels or other circumstances – like bans. Unlike displacement and eviction, banishment has the advantage of capturing both the process of removing the poor *and* of preventing them from coming back. Banishment may not have immediate tangible effects, but it forces the banned to consider whether they are worthy of being in a place and whether it is worth fighting for it. In other words, it leaves the subject of the ban in a permanent state of spatial precariousness. As Roy (2017) formulates it, banishment 'is predicated on the permanently insecure possession of property and personhood'.

Banishment, as opposed to the term 'exclusion' for example, underlines the coercive power of the state to impose spatial segregation. Exclusion often refers to informal power practices that make people unwelcome or unwanted in certain places (Beckett and Herbert, 2009). Unlike the concepts of 'displacement pressure' and 'exclusionary displacement', banishment underscores the coercive power of the state to impose spatial segregation. In Landskrona, banishment is a proactive state policy to alter the socio-demographic fabric of the city (people who want to spend more than a third of their disposable income on rent are not allowed in). Displacement pressure and exclusionary displacement are not imposed policy measures but the indirect outcome of rising housing prices or rent levels.

Banishment is an old phenomenon, but what is new according to Roy (2017) is the renewal of banishment at the urban scale. We witness the (re-) introduction of all sorts of spatial bans in American cities, from park exclusion orders to civil gang injunctions. Based on case studies in Los Angeles and Chicago, Roy (2017) insists that the eviction wave following the foreclosure crisis in the United States should best be interpreted as 'racial banishment' rather than a narrow, legal event – thereby widening the narrow interpretation of eviction as a legal removal act. Roy (2017) insists that the systematic eviction of African Americans from central locations in American cities cannot simply be explained through the sanitized notion of 'displacement', since that leaves open the possibility to interpret the systematic disappearance of African-Americans from city centres as the unfortunate outcome of the – otherwise normal – workings of the housing market. Likewise, 'gentrification' insufficiently underlines the workings of the 'racial state' (Wyly et al., 2012; see also Goldberg, 2001) in the production of a new 'racial-demographic fabric' in inner cities, over and above possible socio-economic transformations in urban neighbourhoods. In the Landskrona case, the racial nature of the banishment tactics is impossible to judge since no statistical evidence is available, while mainstream media in Sweden conventionally do not mention peoples' ethnic or national backgrounds. However, anecdotal evidence (see endnote 1) suggests that even in Landskrona those exclusionary housing policies have a clear ethnic dimension.

Roy (2017) takes her cue from Willie Fleming from the Chicago Anti-Eviction Campaign. In an interview with the UCLA Institute on Inequality and Democracy, Fleming (cited in Vera, 2016) explains why he prefers the term 'banishment' rather than 'displacement' to understand the outcome of the foreclosure crisis in poor Black neighbourhoods:

> Displacement is when the public housing is uprooted and people are moved downtown. … Banishment is when there is no place for you to go. Places for you to go are jails or death. That's banishment. When you look at the affordable housing rates. When you look at affordability on the lowest ends and you see it's not existent. When you look at the bulldozers that are coming in. When you look at the pressures on public housing. You quickly realize that there is nowhere else for you to be.

This quote clearly resonates with the notion of 'city-less citizen' mentioned earlier: under the current banishment regime, poor population segments find it increasingly difficult to be somewhere or go somewhere – they become citizens without a city.

In short, banishment as a concept has the advantage of capturing the discursive exclusion of population segments from certain places, with or without physical consequences; it is better suited to function as the dialectical opposite of boosterism; it captures better the spatial precariousness people may find themselves in; it stresses the proactive role of the state in installing spatial segregation; it allows to highlight the anxieties and temporalities people are exposed to when banned from certain places; and it highlights how the lack or absence of affordable housing 'casts a net' over the city that makes it difficult or impossible for poor population segments to reside in the city. They are banned, even if no such legal order has been issued and even if no authority has told them to move. Rather than an enforceable legal measure, banishment imposes a permanent 'condition' of insecurity and temporality for poor residents that has consequences or one day may have consequences.

The uses of banishment

Through banning poor people, poverty is de facto banned from the city; it does not matter whether or where it continues to exist as long as it does not (re)enter the city. In the process, the social problem of urban poverty is being reduced to a spatial problem – this is what Smith (1993) has labelled 'spatial fetishism'. The use of banishment is so effective because it simply removes poverty from the political agenda: if it is not present, it does not need to be fought. The regulation of space through bans becomes a very powerful tool to set the political agenda. Bans, then, are a very elegant and effective spatial tool to depoliticize a wide range of social problems such as poverty, addiction, mental health, homelessness, begging, sleeping rough, and petty crime, amongst other issues. Even worse, bans lay the blame for socio-economic conditions with the victims: it is the very presence of the poor themselves that is the cause of (inner-city) poverty, and thus, they must be kept out.

Banishment ultimately does nothing to fight the actual problem, implement change in the neighbourhood, take care of the affected, or create awareness around an urban issue. Moreover, it not only fails to address poverty and its associated social problems, it also diverts political attention away from the actual problem: the political problem is not poverty but poor people entering the city. Consequently, it legitimizes a reductionist political agenda that, however superficial, can easily be mobilized in an electoral context, for example. Most crucially, by reducing the problem of poverty to the whereabouts of the poor, the precarious situation of the poor will of course only worsen, since their problems are not addressed properly, or at all.

Banishment may have been (and still is) a flawed tactic to tackle urban poverty, but according to Coy (2008), the banishment of outcasts fulfils deeper functions than merely the removal of bodies. It seeks to regulate and remake society by displaying the limits of acceptable behaviour and asserting civic norms and communal mores, such as work morale. It effectively demonstrates both the power of city authorities and the boundary between inclusion and exclusion in the urban community. Seen this way, banishment is an instrument not just to exclude the poor but also to install social discipline and assert morality over the remaining society. Therefore, the performance of acts of banishment is just as much demonstrations of inclusion as of exclusion.

Banishment is particularly hard to fight since bans tend to be characterized by a considerable dose of hybridity (Beckett and Herbert, 2010). They remain largely invisible to the public, they do not necessitate the use of force, and they are difficult to prove ineffective. Because of their elusive nature, these 'social control mechanisms provide minimal avenues for contestation, thereby diminishing the rights-bearing capacity of their targets' (Beckett and Herbert, 2010: 4). Further, due to the relatively low enforceability of bans, people may not feel the need to organize themselves and fight the ban, if they could, since they can somehow be bypassed or simply ignored at a certain risk.

Further, a ban can be imposed on the poor, but it can also be imposed on landlords who rent out to poor people. This is obviously the case in Landskrona, but it has historical antecedents. There is no space for an overview here, but it suffices to point to at least one historical example with clear analogies with our empirical case. In the second half of the 1500s, a time of warfare and economic hardship in large parts of Europe, many cities were facing increasing immigration from the countryside. The reaction of many city authorities was to drive these unemployed strangers, most often without (serious) criminal record, out of town through banishment. Not that banishment was a particularly effective method to deal with poverty: the penniless poor would often make it back into town – the only place to find food and work. For example, city authorities in the German city of Ulm, failing to staunch the flow of vagrants into the city, issued a new edict in 1586. It not only prohibited the idle from residing inside the city walls but also forbade any citizen to provide housing or charity to any 'discharged soldiers, vagrants, sturdy beggars, and all other dissolute, loose, and good-for-nothing riff-raff' on penalty of a two Gulden fine (Coy, 2008).

In short, banishment is a crude but effective tool because it removes the responsibility of the city authorities for the well-being of poor people, it preserves the status quo for society while making things worse for those affected by the ban, it reminds the rest of society of whom qualifies as a 'worthy citizen', and it is very difficult to resist or mobilize against.

Conclusion

Landskrona is the first Swedish city to implement city-wide policies that prohibit private landlords from renting out to low-income groups. It is not uncommon that landlords implement income requirements, but as a city-wide policy measure imposed upon every landlord, private or public, it is new. This outspoken strategy of the city of Landskrona since 2012 to target people on social benefits or low income in Landskrona illustrates the city's will to change its social fabric not through attracting wealth but banishing want. By halting potential newcomers on benefits or low income and 'unserious' landlords prepared to rent out to them, the city has installed a two-sided process of banishment. On the one hand, banishment is a top-down, city-wide urban renewal strategy that 'casts a net' over the city that closes in on property and personhood, on the right to reside somewhere, without necessary physical displacement or eviction or other immediate action. On the other hand, it is a bottom-up approach that 'pulls the mat' from underneath the feet of the poor, who have nowhere left to reside in the city, by closing in on their housing opportunities and possibilities. The poor effectively become 'citizens without a city'. In either case, the concept highlights the proactive role of the state in establishing a banishment regime in the city that not only removes people but also prevents them from moving in.

Accordingly, banishment can be regarded as a 'bundle of tactics' that can be mobilized at will and imposed in hard or soft ways to curb citizens' residential freedom and landlords' discretionary powers. It can basically take any form, from discursive stigmatization to outright displacement. Whatever the effective outcome, banishment imposes a permanent state of anxiety, insecurity, and temporality on poor residents that may linger for a long time. This has major methodological consequences. In order to capture the full effect of banishment (and its transformation in time, perhaps into new forms of displacement), we need longitudinal research to fully understand its long-term effects for citizens – and for landlords for that matter – and the shifting social-demography of the city.

Note

1 See, for example, the article series in the local newspaper, *Helsingborgs Dagbladet*: Whrethov (2013a, 2013b, 2014) and Petterson (2013).

References

Armstrong, M. F. (1963). Banishment: Cruel and unusual punishment. *University of Pennsylvania Law Review*, 111, pp. 758–786.

Baeten, G. and Listerborn, C. (2015). Renewing urban renewal in Landskrona, Sweden: Pursuing displacement through housing policies. *Geografiska Annaler B*, 97(3), pp. 249–261.

Beckett, K. and Herbert, S. (2009). *Banishment. The new social control in urban America*. Oxford: Oxford University Press.
Beckett, K. and Herbert, S. (2010). Penal boundaries: Banishment and the expansion of punishment. *Law & Social Inquiry*, 35(1), pp. 1–38.
Beckett, K. and Herbert, S. (2017). Banishment and the post-industrial city: Lessons from Seattle. *European Journal on Criminal Policy and Research*, 23(1), pp. 27–40.
Blomley, N. (2004). *Unsettling the city: Urban land and the politics of property*. London: Routledge.
Bosman, J. and Barbaro, M. (2009). Mayor defends one-way tickets for homeless. *The New York Times*, 29th of July, 2009. https://cityroom.blogs.nytimes.com/2009/07/29/mayor-defends-one-way-tickets-for-homeless/ [Accessed 29 June 2020].
Brant, M. (2011). Vill flytta bidragstagare och riva hus. *Landskrona Posten*, 24 January 2011.
Brant, M. (2019). Lista: Sprängningar och skjutningar i Landskrona. *Helsingborgs Dagbladet*, 8 August 2019.
Coldham, P. W. (1992). *Emigrants in chains*. Baltimore, MD: Genealogical Publishing Company.
Coy, J. P. (2008). *Strangers and misfits: Banishment, social control, and authority in early modern Germany*. Leiden: Brill.
Goldberg, T. D. (2001). *The racial state*. Malden, MA: Blackwell.
Gutsche, R. (2015). Boosterism as banishment: Identifying the power function of local, business news and coverage of city spaces. *Journalism Studies*, 16(4), pp. 497–512.
Hartman, C. W., Keating, W. D., and LeGates R. T. (1982). *Displacement, how to fight it*. Berkeley, CA: National Housing Law Project.
Landskrona Stad. (2012). *Landskronas vägval. Centrum/Öster*. Landskrona.
Landskrona Stadsutveckling. (2017). *Årsredovisning 2017*.Landskrona.
Lees, L., Slater, T. and Wyly, E. (2008). *Gentrification*. New York: Routledge.
Listerborn, C. and Baeten, G. (2016). 'Landskronas vägval': När bostäder åt alla blev ett kommunalt problem. In: Ek, R. and Tesfahuney, M. (eds.) *Den postpolitiska staden*. Sandared: Recito Förlag, pp. 52–80.
Lönnaeus, O. (2012). Staden som vill bli som alla andra. *Sydsvenskan*, 23 September 2012. Available at: https://www.sydsvenskan.se/2012-09-23/staden-som-vill-bli-som-andra
Marcuse, P. (1985). Gentrification, abandonment, and displacement: Connections, causes, and policy responses in New York City. *Journal of Urban and Contemporary Law*, 28, pp. 195–240.
Milne, R. (2018). Denmark proposes doubling punishment for crimes in 'ghettos'. *Financial Times*, 1 March 2018.
Mitchell, D. (1997). The annihilation of space by law: The roots and implications of anti-homeless laws in the United States. *Antipode*, 29(3), pp. 303–335.
National Board of Housing, Building and Planning. (2014). *Urbant utvecklingsarbete – delrapportering nr. två av regeringsuppdrag*, Rapport 2014:9, Karlskrona.
Petterson, P. (2013). Är du välkommen in Landskrona? *Helsingborgs Dagbladet*, 16 September 2013.
Petterson, P. (2014). Minskad bidragsinflyttning glädjer Strandberg. *Helsingsborgs Dagbladet*, 1 June 2014.
Roy, A. (2017). Dis/possessive collectivism: Property and persoonhood at city's end. *Geoforum*, 80, pp. A1–A11.

SCB. (2018). *Statistik: skattekraft*. Stockholm: SCB.

Simonsen, K. B. (2016). Ghetto – Society – Problem: A discourse analysis of nationalist othering. *Studies in Ethnicity and Nationalism*, 16(1), pp. 83–99.

Smith, N. (1993). Geography, difference and the politics of scale. In: Bird, J. et al. (eds.) *Mapping the Futures. Local Cultures, Global Change*. London: Routledge, pp. 55–79.

Socialstyrelsen. (2018). *Statistiskdatabas: ekonomisk bistånd*. Stockholm.

Swedish National Housing Board (Boverket). (2015). *Uppföljning av det prestationsbaserade stimulansbidraget – delrapport 2*, Rapport 2015:17.

van Gent, W., Hochstenbach, C., and Uitermark, J. (2017). Exclusion as urban policy: The Dutch 'Act on Extraordinary Measures for Urban Problems'. *Urban Studies*, 55(11), pp. 2337–2353.

Vaver, A. (2011). *Bound with an Iron Chain: The untold story of how the British transported 50,000 convicts to colonial America*. Westborough: Pickpocket Publishing.

Vera, C. (2016). Against gentrification, against banishment: An activist toolkit (from Willie 'JR' Fleming) to keeping hoods yours. *Institute Graduate Student Researchers Interview Urban Color-Lines Activists*. [Online]. Institute on Inequality and Democracy. Available at: https://challengeinequality.luskin.ucla.edu/2016/03/02/luskin-school-students-interview-urban-color-lines-activists/#second

Wacquant, L. (2009). *Punishing the poor. The Neoliberal Government of Social Insecurity*. Durham, NC: Duke University Press.

Wacquant, L., Slater, T., and Pereira, V. B. (2014). Territorial stigmatization in action. *Environment and Planning A*, *46*(6), pp. 1270–1280.

Whrethov, E. (2013a). Jag har fått nej på alla ansökningarna. *Helsingborgs Dagbladet*, 13 September 2013.

Whrethov, E. (2013b). Han fick en bostad till slut. *Helsingborgs Dagbladet*, 4 November 2013.

Whrethov, E. (2014) Socialen får backa om tvångsflytt. *Helsingborgs Dagbladet*, 15 March 2014.

Wyly, E., Ponder, C. S., Nettling, P., Ho, B., Fung, S. E., Liebowitz, Z., and Hammel, D. (2012). New racial meanings of housing in America. *American Quarterly*, 64(3), pp. 571–604.

8 Common resistance against state-led stigmatization and displacement

Louise Fabian and Anders Lund Hansen[1]

Introduction

'Housing is a human right' reads a sign carried by a young woman at the Almen Modstand (Common Resistance) demonstration in Copenhagen, 29 September 2018. The movement was created earlier the same year as a reaction to the commodification, privatization, and territorial stigmatization of the common, non-profit housing (*almene boliger*) in general and to the government's 'ghetto plan' in particular that are leading to the displacement of thousands of peoples. Coordinated parallel demonstrations organized by Almen Modstand took place in two other major cities in Denmark: Aarhus and Odense (see Figure 8.1). Based on an analysis of the Danish state's production of inequality through housing policy, we explore state-led stigmatization and displacement in a Nordic (post)welfare context (Baeten et al., 2015) – where 'displacement can take very subtle (yet effective) forms' (Beaten et al., 2020, Introduction to this volume). Moreover, we examine the new forms of (common) resistance to the assault on common, non-profit housing and marginalized peoples.

'Losing one's place can be much more traumatic than simply changing location', explains Smith (1994: 253–254). Research on displacement and dispossession shows how these acts of violence are commonly sanctioned by the state and inflicted upon undesired minorities and marginalized low-income communities (e.g., Atkinson, 2000; Hartman and Robinson, 2003; Fullilove, 2004; Lees et al., 2008; Porter, 2009). Additionally, when studying processes of marginalization, Wacquant (2016) suggests that we should look into the institutional mechanisms and policies that produce, reproduce, and change networks of position and racialized marginality and stigmatization.

This chapter focuses on the role of the Danish state in stigmatizing and displacing these groups from targeted common, non-profit housing estates (the so-called ghettos). This politics has spurred critique and housing activism around themes related to 'the right to the city', 'housing as a human right', and 'right to freedom from displacement' – issues clearly linked to transnational discussions (e.g., Attoh, 2011; Fitzpatrick et al., 2014; Hartman 2002; Lund Hansen and Clark, 2017). In this chapter, we look closer at these resistance

Figure 8.1 Almen Modstand (Common Resistance) demonstration, Copenhagen, 29 September 2018. Text on sign: Housing is a human right.
Photo: Anders Lund Hansen.

movements, with special focus on Denmark's largest non-profit housing estate, Gellerupparken. We draw on an intersection between the literature of critical urban theory and social movement research (Fabian et al., 2015; Mayer, 2016; Fabian and Nielsen, 2020). This perspective encourages understanding recent urban social movements in the context of general socio-economic processes behind the production of contemporary cities. Accordingly, we conducted an analysis of the neoliberalization of Danish housing (Larsen and Lund Hansen, 2015), and our findings suggest that the stigmatization of underprivileged areas and peoples and the displacement of precarious groups are intrinsically linked to forms of neoliberal urbanism (e.g., privatization of common housing). Our study thus suggests that the ensuing resistance stems from different but intertwined political processes.

The study builds partly on earlier research into related processes and partly on novel material collected during the past years. The material has been collected through studies of policy documents (in particular, the shifting so-called ghettopackages, government strategies, the architectural masterplan Helhedsplan Gellerup, and New Year's Speeches by Danish prime ministers); around fifty formal and informal interviews and group interviews with central actors (involved in urban planning, housing organizations, and resistance activism); participant observations in public meetings and panel debates; and our own organic intellectual commitment to social change.

We start this chapter with a background section, where we highlight the increasing production of inequalities in Denmark and the concentration of precarious socio-economic groups in rental housing. We then scrutinize the recent history of the non-profit housing sector and the production of discontent through attempts to dismantle and weaken the non-profit housing sector. Subsequently, we explore the government's ghetto politics, relevant policy documents, and state-led territorial stigmatization and displacement. We then address how different actors have been trying to subvert the ghetto stigmatization through different forms of resistance – such as the new social movement Almen Modstand (Common Resistance) as well as independent activists fighting against what they see as a racist politics. Finally, we conclude that stigmatization of underprivileged areas and peoples and the displacement of precarious groups is a product of different (but intertwined) economic and political processes. We also discuss the extent to which new forms of activism can reinvigorate discussions related to race, class, and gender tensions in contemporary housing questions.

Background: producing inequality and discontent through housing politics

Constituting twenty per cent of the total Danish housing stock, non-profit housing has traditionally been seen as a solution to the housing question and a cornerstone in the Danish welfare state. However, this model is currently under attack (Larsen and Lund Hansen, 2015; Fabian et al., 2019). Parallel to, for example, some of the marginalized French *banlieues* or the Swedish Million Programme estates, typically built in the boom period after the Second World War, many large-scale housing estates in Denmark are also stigmatized as 'ghettos'. In the latest assault, the Danish government announced a plan 'to terminate the ghettos completely', as the then Prime Minister Lars Løkke Rasmussen stated in his 2018 New Year television address to the Danish people. In some housing areas, the idea is to '[t]ear down buildings. Spread the inhabitants and rehouse them in different areas'. In other areas, the government wants to 'control who moves in'. Thus, the overall goal is to 'close the gaps in the map of Denmark' (Rasmussen, 2018, own translation). The Government's strategy is named 'Et Danmark uden parallelsamfund – ingen gettoer i 2030' (A Denmark without parallel societies – no ghettos in 2030). The strategy was launched on 1 March 2018 by the Danish prime minister and seven ministers in a highly staged event in the Common Housing estate Mjølnerparken in Copenhagen – an area known for its high percentage of tenants with an immigrant background. Before discussing the current attacks on the sector in general and the displacement and stigmatization of the most socio-economically precarious parts of the population in particular, we will take a closer look at the social-geographical conditions in Denmark and the non-profit housing sector.

Historically, there have been various political solutions to the Danish housing question. The rent-seeking behaviour of finance capital and landed developer interests generally drives the formation of market relations through privatization and commodification of built and natural environments (Larsen et al., 2016). Danish housing markets are no exception. Neoliberal reforms have caused increased socio-geographical inequalities and a concentration of socio-economically precarious groups in rental housing – especially in confined common, non-profit housing areas. These areas are now under attack from, amongst others, the government's 'ghetto plan', which is stigmatizing and displacing people to make way for commodification, privatization, and White middle-classification.

As Piketty (2014) demonstrates, income inequality has increased significantly in rich countries over the past decades. This is also true for Denmark (see Figure 8.2) and other Nordic countries – traditionally known for their equality – as they are exercising their own '"messy hybrid" of neoliberalizing local and national politics' (Mayer, 2016: 61).

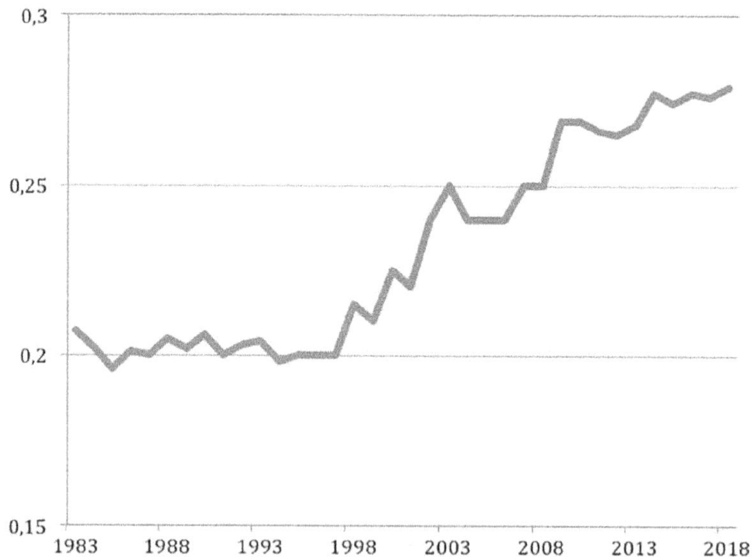

Figure 8.2 Gini coefficient in Denmark, 1983–2018.
Source: Eurostat and Danish Ministry of Finance.

According to the Danish Ministry of Finance (2004) and Ministry for Business and Growth (2013), the striking changes that began at the end of the 1990s and took off in the 2000s are rooted in the increased capital gains in private housing and stock markets. (For a more detailed account of the events, see Lund Hansen et al., 2015.) This increasing polarization has a geographical dimension. In the past decades, the four largest urban regions in Denmark experienced increased segregation based on income, education, and individuals on social benefits. This tendency was most severe in Greater Copenhagen and Aarhus (Ministry of Housing, Urban, and Rural Affairs, 2014; AE, 2018a, 2018b; Eurostat, 2018). Most of the low-income groups in Denmark rent their homes (Statistics Denmark, 2014). Therefore, below we will discuss the changes in Denmark's largest rental sector: non-profit housing.

Non-profit housing (or in Danish *almene boliger*) is a form of Scandinavian common housing. The sector stands out in a number of ways, especially its multi-scalar, democratic organization. The sector comprises approximately 800 housing associations and 8,000 branches, with a total of more than half a million units. It provides shelter to nearly one-fifth of the population; thus, it is still a cornerstone in the Danish welfare state. Organized in independent housing associations, non-profit housing can be considered 'collective private property'. The sector's well-established, multi-scalar tenants' democracy is one of the 'distinctive features' of Danish non-profit housing, also in a Nordic context (Madsen, 2006: 33).

The vast majority of units were built in the post-1945 building boom, and like in many other European cities, they are mainly located in and around the large cities (Andersen, 2012). What started out as utopian modernist solutions to the housing question became in the 1970s and 1980s perceived as dystopian mass housing for socio-economically marginalized groups (Schierup, 1994; Høghøj, 2019). The housing sector has traditionally accommodated a diversity of socio-economic groups, but since the 2000s, there has been an overrepresentation of low-income, unemployed, immigrant, one-person, and single-parent households living in non-profit housing (Ministry of Social Affairs, 2006).

Traditionally, the sector has received public support from the state and municipalities for construction and repayment of loans. Since 1947, the municipalities have the right of disposition over one-quarter of the sector's housing units to provide shelter for people in need. Historically, the local and the central state have supported both 'bricks' (direct subsidies to construction) and 'people' (housing allowances). However, there are subsidies to all housing sectors, and indirect subsidies like tax deduction to owner-occupied and cooperative housing are greater than the subsidies given to the non-profit sector (Erlandsen et al., 2006).

Leading up to the government's current ghetto plans, at least three main attempts to weaken the non-profit housing sector occurred during the 2000s. (For a more detailed account of the events, see Larsen and Lund Hansen, 2015.) The first strike was the appropriation of the sector's collective savings,

Landsbyggefonden (The national building fund) – a manoeuvre that has contributed to a general 'undermining of the sector's institutional platform' (Jensen, 2013: 59), thereby potentially paving the way for a dismantling and commodification of the sector. The second strike came in 2004 with the introduction of a Danish version of Thatcher's 'right to buy' scheme. The (neo) liberal mayor for construction and engineering in Copenhagen, Søren Pind, called the legislation 'a battering ram straight in the heart of the non-profit housing movement' (cited in Møller, 2002). Due to a strong housing sector at the time, the attempt was not 'successful'. The case

> reveals the contradictory realities of Danish non-profit housing: because of the private (if collectively) [sic] nature of the association-based model, non-profit housing associations seemed – rather paradoxically – able to resist attack from a government hailing the neoliberal virtue of private property.
>
> (Larsen and Lund Hansen, 2015: 270)

The residents' local-regional-national democracy is a special feature of the non-profit sector in Denmark; it is unique internationally, and it played an important role stopping the privatization of the sector. Hence, it is hardly accidental that the third strike in 2009 – in the form of a (new public) management reform – was aimed at weakening the local democracy. The events exposed a conflict of interest in 'the sector' between the non-profit housing associations and the local state, on one side, and its tenants on the other (LLO, 2008). This conflict is not new, yet it is clear that the law is designed to exercise top-down management of marginalized and stigmatized groups such as 'poorly integrated' immigrants – a group under constant attack since the 2000s, when the xenophobic People's Party became a key part of the right-wing government. With these events in mind, it is hardly surprising that the group Almen Modstand sees the light of day a decade later. Before we move on to this novel social movement, we zoom in on the triggering initiatives launched by the Danish government.

State-led territorial stigmatization and displacement

As previously mentioned, the Danish government's ghetto strategy, launched in March 2018, was called 'A Denmark without parallel societies – no ghettos in 2030'. The title of the strategy clearly states that its official target is to make Denmark completely free from so-called ghettos by 2030. However, discourse analysis of the strategy shows that concepts such as 'ghettos' and 'parallel society' are used and referred to in the strategy as indisputable facts, when in fact they are state-sanctioned discursive constructs that are establishing what they are naming. The strategy itself creates 'ghettos' and 'parallel society'.

The discursive ghetto politics was in fact already used by the former Danish Prime Minister Anders Fogh Rasmussen in his New Year's speech in 2004, and later again the same year when the then government launched its 'Strategi mod ghettoisering' (Strategy against ghettoization). In 2005, it was decided that every year, the Ministry of Social Affairs should publish a list of the current 'ghettos'. Being on the list legitimizes specific politics and laws being conducted in the designated place. When the right-wing Danish government re-launched the ghetto debate and published its 'ghetto plan' on 26 October 2010 with the new Prime Minister Lars Løkke Rasmussen, it was called 'Ghettoen tilbage til samfundet. Et opgør med parallelsamfundet i Danmark' (Return of the Ghetto to Society: Taking Action against Parallel Societies in Denmark). The Danish ghettoes were from then on defined as areas where (1) at least fifty per cent of the inhabitants are immigrants or children of immigrants, (2) at least forty per cent of the inhabitants are without connection to either education or the labour market, and (3) at least twenty-seven individuals are convicted of crime per 1,000 inhabitants. Notably, it was not only the right-wing government who, at this point in time, used the term ghetto. In the same year – 2010 – the Social Democrats and the Socialist People's Party also presented a joint ghetto initiative. Since June 2019, when the Social Democrats came into power, they have continued the politics.

The word *ghetto* is traditionally used to refer to crowded urban quarters of a city where a minority lives – often under social, financial, and political pressure. Ghetto was originally the name of an area in Venice where Jews were restricted and segregated in 1516. The different ghettos in Europe where Jews lived in medieval times were often strictly segregated regulated areas, established by societies that repressed the Jews. One of the darkest chapters in the history of ghettos was when the Nazis during the Second World War confined Jews and Romani People in restricted areas of Eastern European cities, mostly referred to as Jewish Quarters. However, when Danish politicians chose to use the term 'ghetto' to refer to areas in Denmark with a high population of people with 'non-Western background', they were drawing on an imaginary from North American ghettos (especially African American ghettos) as pathological spaces of dereliction, violence, poverty, and segregation. Sociologist Loic Wacquant (2004, 2008, 2015) warns that we have to be aware of the fundamental differences between divergent forms and processes of socio-spatial seclusion, whereby specific populations are being isolated willingly or by force. The American ghettos developed in close connection to immigration, where people from different European countries in the mid-nineteenth century gathered in immigrant communities in specific areas. It has been argued that forming ethnic enclaves can help people build support-networks, share knowledge, and grow social capital. Wacquant (2016) makes a distinction between, on the one hand, ethnic enclaves that work as support communities acting as a springboard on the way to cultural assimilation and social mobility and, on the other hand, ghettos that keep people in a more permanent condition of segregation, stigmatization, and material and symbolic isolation, with limited economical and social possibilities. Wacquant (2018) teaches us how, when studying processes of marginalization, we should look into the institutional mechanisms and policies that produce, reproduce, and change networks of position and racialized marginality.

But what is it that the Danish government is doing when it chooses to use the label ‘ghetto’ to officially define areas based on the amount of people with immigrant background, unemployment status, and conviction records? The word *ghetto*, together with the term *parallel society*, is used as a political category in the Danish debate in order to name and claim ‘the ghetto’ as an antagonistic unadaptable Other – an anti-identity that cannot in its current state be integrated into Danish Society. The discursive construction of the ghetto is used to target immigrant and poor people living in these areas, especially people with a Muslim background; these people are constructed as a negation of democracy and a negation of what is explicitly called ‘Danish values’ in the policy documents.

Where historically ghettos are characterized by being inhabited by populations with a relatively similar ethnic background, the Danish so-called ghettos consists in reality of a rich variety of nationalities, including a large amount of ethnic White people, mainly people with a low income. For instance, Gellerupparken – our main case study in the chapter – has inhabitants from over eighty nationalities.

Gellerup is an area in the outskirts of Aarhus that is dominated by non-profit housing. Eighty per cent of the inhabitants in the housing project Gellerupparken are of non-Danish descent. Gellerupparken – the largest non-profit housing area in Denmark currently undergoing a massive redevelopment – was built between 1968 and 1972. Before the demolition of high-rise concrete buildings in the area started, it contained 2,400 flats and over 7,000 people. Over eighty per cent of the people living in the area are immigrants or descendants of immigrants, and over fifty per cent are unemployed or lack educational affiliation. It has figured on the government’s ghetto list as a ‘hard ghetto’ since the list’s inception. In 2011, the most expensive high-rise renewal masterplan in Danish history was approved. It has a budget of DKK 6.5 billion. The official goal of the plan, called Helhedsplanen, is to make Gellerup multifunctional: ‘for the diversity of the community here to match the rest of Aarhus in terms of employment, level of education, income and safety’ (Helhedsplan Gellerup, 2018). In reality – and this is confirmed by interviews we have made with both initiators of the plan and architects conducting the plan – the goal is to attract more ‘resourcefull’ inhabitants with a non-immigrant background. Besides selling land and buildings to private companies and investors and demolishing nine blocks, the plans involves a number of ways to try to make physical solutions to socio-economic problems. An architect involved in the project characterizes the transformations as ‘having Haussmannian dimensions’ (Anonymous Architect, 2019).

Figure 8.3 "The golden gate". Reconstruction in Gellerupparken, 2019.
Photo: Louise Fabian.

As seen in Figure 8.3 a huge golden hole has been made in one of the blocks in order create a gate and lead traffic into the area, a symbolic opening up to the rest of the city and a symbolic demonstration of the claimed fight against the formation of 'isolated parallel societies'. The hole in the building block was created by demolishing seventeen homes and, consequently, displacing the families living there – a prelude to the thousands of people that will be directly displaced when all the plans are realized (Gellerup Planen, 2018; Hansen, 2019). When Gellerupparken was originally built, it was purposely built with very few roads in order to create safe zones for kids and pedestrians. The large amount of infrastructure and roads now created aims to make the area accessible for people from other parts of town. A new broad boulevard is being created through the area, named Blixen Boulevard after the Danish author Karen

Blixen. Her indisputable literary and human qualities aside, Blixen was not only an author, she was also a White, privileged aristocrat running a farm in colonial (pre-independence) Kenya. It is hard to ignore the irony that at the same time as it is being discussed in Kenya to remove the name Karen from a suburb in the capital, Nairobi, because it presumably honours an author who had a number of racist formulations and ideas – for instance, in her memoir *Skygger på græsset* (Shadows on the grass, 1969), she compared Africans to children – the name is used in a Danish project aimed at demonstratively 'colonizing' and reappropriating an area presently dominated by immigrants from a non-Western background and replacing them with more resourceful and in effect more White inhabitants (Figure 8.4).

Figure 8.4 Constructions in Gellerupparken, 2019, and the new road sign.
Photo: Louise Fabian.

Subverting ghetto stigmatization and common resistance

When we look into the present form of resistance toward the 'ghetto package', we see not only how the Danish government used word *ghetto* to construct a negative identity that politicians can use to try to displace unwanted citizens but also how the word has been used by people protesting the displacement to challenge, mock, and reinterpret the very term *ghetto*. Some people living in the designated areas have critically adapted the imposed term to subvert the identification; they use mocking and playing with described negative identities as a way of creating agency, political subjectivity, and solidarity. For instance, the art gallery Andromeda in Gellerupparken has organized the project Demolition Tours with the guide Aysha Amin in a bulletproof vest (see Figure 8.5) in order to resist the process of territorial stigmatization (Kallin and Slater, 2014; Wacquant et al., 2014), to facilitate embodied cross-cultural encounters (Simonsen and Koefoed, 2020) and to mock the narrative of the ghetto being a non-integrated area in a condition of violence and social antagonism.

Andromeda has had a strong agenda of creating alternative representations of Gellerup from the residents' perspective, for example, through the production of zines and a series of t-shirts with photographs of the buildings being threatened by demolition. One of these self-produced zines is called

Figure 8.5 Demolition Tours in Gellerup, 2018, with tour guide Aysha Amin from Andromeda.
Photo: Mads Christiansen.

'Decolonise Your Tastebuds' and was made by a small group of residents in three days, the last day concluding over a dinner table with a Somali dish of kidney and vegetables. In the zine 'Decolonise Your Tastebuds', the recipe for the dish is enclosed together with reflections like 'When I eat pasta, I feel like I eat the colonizers food' and 'turning food into politics'. Aysha Amin explains how one of the ideas behind creating these zines is 'to reclaim media representations and create alternative spaces for discussion around culture and the ongoing transformations' (Amin, 2019). Notably, the activism of Andromeda addresses both external and internal processes of colonization and aims at empowering, connecting, and 'decolonizing people's minds' (see Figure 8.6).

Our interviews with adult children of immigrants (who wanted to be anonymous) revealed how this group experiences stigmatization and racist politics as something that is so unfair and discriminating that it makes

Figure 8.6 Zine made by Andromeda.
Photo: Louise Fabian.

them want to give up the idea of ever being accepted as Danish citizens. Thus, the politics that claim to have integration as its goal in effect functions as a process of Othering that isolates and displaces, both physically and mentally. Helle Hansen, an ethnic Danish citizen who has been involved in Afdelingsbestyrelsen (the local board of the housing association) since 2002, and from 2007 to 2016 as head of the board, also expresses a feeling of deep frustration and isolation: 'I feel radicalized and outcasted' (Hansen, 2019).

Immigration has been a hot and highly symbolic topic in Danish politics since the 1990s, and parties that adopt an immigrant-sceptical politics have been steadily growing since – resulting in ever-stricter immigration politics and ever-harder rhetoric. Immigrants are portrayed as a security and stability threat and as an economical and ideological burden to the welfare system. In the rhetoric of the 2018 'ghetto plan', we clearly see a construction of a group of people who does not support 'Danish values'. The government's 2018 'ghetto plan' states, 'It is a threat to our modern society, when freedom, democracy, equality, and tolerance are not accepted as founding values'.

The latest Danish Ghetto Plan, formulated in March 2018 by Denmark's right-wing coalition government, claims to combat social problems in twenty-five neighbourhoods considered particularly vulnerable and problematic. As already mentioned, a neighbourhood classifies as a ghetto depending on the number of unemployed living in the area, crime statistics, and most importantly, the proportion of inhabitants with immigrant background. The original package that was tabled in Parliament in March 2018 included twenty-two bills. A great number of the suggested bills has since then been negotiated and supported by the right-wing coalition government, the right-wing Danish People's Party, and two opposition parties: the Socialist People's Party and the Social Democrats. Kaare Dybvad, spokesperson for the Social Democrats, has argued that the party supports the politics because it is convinced that concentrating social problems increases social problems. The plan has a number of laws that affect the non-profit housing sector and control who can move in and out. The plan also consists of a number of laws that specifically target immigrants and children of immigrants from a non-Western background. Parents who take their children to their country of origin for long periods of time or parents to children that drop out of school can face cuts in welfare benefits or a prison sentence. Certain crimes committed in ghetto areas, including vandalism and theft, can be punished with penalties twice as hard as those committed elsewhere. These laws break with the fundamental principle of 'equality for the law' that is meant to protect citizens against discrimination, and that is a fundamental part of international human rights' legal framework. As we have shown, the 'ghetto package' is part of an ongoing nationalist agenda, where the right-wing Danish government has made a number of laws that make it unattractive and often impossible for refugees to come to and seek refuge in Denmark. In September 2015, the Danish government placed ad in four Lebanese newspapers to scare away potential would-be migrants from coming to Denmark. The ads informed about how Denmark had tightened the regulations, made significant cuts in social benefits for newcomer refugees and made harsher rules for obtaining the permanent residence permit or getting family reunification.

New social movements have arisen throughout the world around the struggle for affordable housing. In Denmark, in opposition to the 'ghetto plan', we see alliances between different precarious groups of citizens affected by the threat of displacement and demolition of their homes. The 'ghetto package' has been met with a growing number of resistance movements and activities in Denmark. The national grassroots movement Almen Modstand (Common Resistance)[2] has been a main player in mobilizing the public against the 'ghetto package' and its attack on Danish non-profit housing. Almen Modstand is anchored in different housing areas across Denmark, where it has local committees. It has managed to get a fair share of public attention and mobilized both tenants and people sympathetic with its cause. Almen Modstand has four central demands: (1) influence on decision making, (2) a democratic process, (3) an end to the selling of and speculation in non-profit housing, and finally (4) the elimination of the 'ghetto list' and its criteria. Almen Modstand organizes meetings, demonstrations, and happenings and educates people and affected tenants about the relevant juridical processes.

Along with conducting participant observations in activities and demonstrations organized by Almen Modstand, we have worked together with and taken part in a number of activities organized by the small art gallery Andromeda in Gellerupparken. As previously mentioned, Andromeda has worked as a platform for workshops, artistic activities, and activism that challenge the ghetto politics and discourse. The gallery tries to tell a different story about the place and to fight against the plans to demolish thousands of homes in the area.

The 'home' as well as the 'community' are both lived experiences, imagined ideals, and potential arenas of collective action and experience (Hooks, 2015; Fabian, 2017; Jupp, 2017). The domestic sphere is traditionally considered private, but the question of housing brings the private to the public. Affordable and available housing is a matter of general political concern, and Almen Modstand mobilizes a great number of people from marginalized communities who have so far not been politically active; solidarity and resistance arise around common threats to their homes. Meetings, demonstrations, and online activity become arenas for collective action and identification (Fabian and Reestorf, 2015).

The attack on non-profit housing in Denmark has come from a broad coalition of political parties. As shown in this chapter, it has been conducted by a politics of Othering those being governed by the 'ghetto plan' – people with an immigrant background as well as people dependent on social benefits. Neoliberal housing politics works here in tandem with xenophobic politics. Economic disinvestments in the sector, neoliberal reforms, and a general culture of blaming and stigmatizing those in need of support result in isolating the people living in places like Gellerup. Austerity hits at the level of the domestic sphere, the body, the family, and the home (Jupp, 2017). Therefore, as Fabian et al. (2019) argue, resistance often tries to mobilize and generate the exact opposite logics and generate pride, solidarity, and cooperation.

Denmark is a striking example of how heralding of mixed-income housing and poverty deconcentration are being used to demolish public housing and replace it with private property. Despite the rhetoric around improved integration and celebration of a 'tamed multicultural city', there is a great chance that the current Danish politics of state-led territorial stigmatization and housing commodification will in reality lead to increasing socioeconomic division, increased polarization, and dynamics of social exclusion and displacement.

Conclusions

In our study of state-led stigmatization and displacement, we have seen a trifocal attack against undesired people and places. The first was the neoliberal 'battering ram straight in the heart of the Danish non-profit housing movement', as Copenhagen mayor Søren Pind, supporter of the attack, 'poetically' called the early attempt to commodify the sector. The second is the stigmatization and radical Othering found in the discourses of the different ghetto plans, which position 'the ghetto' in a pathological state of antagonistic Other, not adaptable to 'Danish values' and society. The third is the physical battering ram of urbicide (Schneider and Susser, 2003) and renoviction (Molina and Westin, 2012) of housing projects such as Gellerupparken, opening the areas for the White, (upper)middle classes – and eventually displacing thousands of people.

The attack on the non-profit housing in Denmark has been supported by a broad coalition of political parties through a politics of territorial stigmatization and Othering of people with an immigrant background as well as poor people on social benefits. The discourse of 'parallel society' blames and stigmatizes vulnerable citizens; it individualizes, and it separates. Nevertheless, the forms of resistance mentioned in this chapter – namely, Almen Modstand and Andromeda – oppose this politics of blame and stigma and try to resist and subvert the 'ghetto' discourse, thereby generating knowledge, solidarity, and alternative futures.

The history of rent strikes and tenant mobilization shows that the power of the tenant stems from solidarity, organization, and mobilization. Repeatedly throughout history, housing activists have allied with labour unions, organizations working for social reform, and political parties. Right now, housing activists seem rather alone in the parliamentary political landscape. Only the Red–Green party Enhedslisten has been openly supportive of and active in Almen Modstand. In a time of neoliberal urban restructuring, housing activists are forming new alliances and exploring new protest repertoires in order to mobilize resistance to eviction, marginalization, and stigmatization – and to fight for the right to freedom from displacement. And for a politics where housing is seen as use value rather than exchange value. It remains to be seen how efficient these tactics will be in subverting race, class, and gender injustices in contemporary housing questions. What remains is a general picture of housing policy as a social-geographical and ideological battlefield where housing is seen as a strategic tool and an investment object rather than a framework for human life. Housing is where we get to meet our basic ecological and social needs. A responsible housing policy should therefore not use housing as a starting point for a politics of displacement and a product in the traditional sense. Rather housing should be realized as a solid base for human being, togetherness, and habiting.

Notes

1 We wish to acknowledge that research underlying this chapter in different ways was supported by the European Commission, HERA, fourth joint research programme: Public Spaces: Culture and Integration in Europe; the FORMAS funded Strong Research Environment: CRUSH – Critical Urban Sustainability Hub; the VELUX foundation funded project: Gender Blender; and the Aarhus University based Urban Orders Research Center (URO).
2 http://almenmodstand.dk.

References

AE. (2018a). *Stigende polarisering i Danmark. Siden krisen har indkomsterne stået stille i dele af Danmark* [Increasing polarization in Denmark. Since the crisis, income has remained stagnant in parts of Denmark]. Copenhagen: Arbejderbevægelsens Erhvervsråd.

AE. (2018b). *Indkomster deler i stigende grad hovedstadsområdet og Århus* [Incomes are increasingly dividing the metropolitan area and Aarhus]. Copenhagen: Arbejderbevægelsens Erhvervsråd.

Amin, A. (2019). *Interview*, Aarhus, 8 April.

Andersen, H. T. (2012). The solidity of urban socio-spatial structures in Copenhagen. In: Maloutas, T. and FujIta, K. (eds.) *Residential segregation in comparative perspective: Making sense of contextual diversity*. Farnham: Ashgate, pp. 177–196.

Anonymous Architect. (2019). *Informal interview*, Aarhus, April 7 2019.

Atkinson, R. (2000). Measuring gentrification and displacement in Greater London. *Urban Studies*, 37(1), pp. 149–165.

Attoh, K. A. (2011). What kind of right is the right to the city? *Progress in Human Geography*, 35(5), pp. 669–685.

Baeten, G., Berg, L. D., and Lund Hansen, A. (2015). Neoliberalism and post-welfare Nordic states in transition. *Geografiska Annaler B*, 97(3), pp. 209–212.

Baeten, G., Listerborn, C., Persdotter, M., and Pull, E. (2020). Introduction: Housing displacement: Conceptual and methodological issues. In: *Beyond displacement: Conceptual and methodological issues.* London: Routledge.

Erlandsen, E., Lundsgaard, J., and Huefner, F. (2006). The Danish housing market: Less subsidy and more flexibly. *OECD Economics Department Working Papers 513.* Paris: Organisation for Economic Co-operation and Development.

Eurostat. (2018). *Eurostat statistical books: Living conditions in Europe.* Luxembourg: European Union.

Fabian, L. and Reestorff, C. M. (2015). Mediatization and transformations of cultural activism. Mediatized cultural activism, special issue. *Conjunctions: Transdisciplinary Journal of Cultural Participation*, 2(1), pp. 1–20.

Fabian, L. (2017). Prekære kroppe mod Trump: Judith Butler læser Trump. In Thorup, M. (ed) Uhørt! Idéhistorien læser Donald Trump. Aarhus: Baggrund, pp. 125–129.

Fabian, L. and Nielsen, L. Y. (2020). Gendering social movements and everyday resistance in conjunctions. *Transdisciplinary Journal of Cultural Participation*, 7(1), pp. 1–22.

Fabian, L., Lund Hansen, A., Jenisch, N., and Kaarup-Christensen, E. (2020). Housing activism, neoliberalization and the right to stay put (accepted for publication).

Fabian, L, Nørgaard, A. E., and Risager, B. S. (2015). Protestens taktikker, traditioner og teorier, redaktionelt forord. *Slagmark – tidsskrift for idéhistorie*, 71(6), pp. 11–27.

Fitzpatrick, S., Bengtsson, B., and Watts, B. (2014). Rights to housing: Reviewing the terrain and exploring a way forward. *Housing, Theory and Society*, 31(4), pp. 447–463.

Fullilove, M. T. (2004). *Root shock: How tearing up city neighborhoods hurt America and what we can do about it.* New York: Ballantine Books.

Gellerup, P. (2018). *Udviklingsplan for Gellerup og Tovesvej 2018. Liv i bydelen: Fremtidens Gellerup.* Aarhus: Brabrand Boligforening.

Hansen, H. (2019). *Interview. Skype, in person and by telephone*, 3, 7 and 8 April.

Hartman, C. (2002). Right to stay put. In: *Between eminence and notoriety.* New Brunswick, NJ: CUPR Press. (Original work published 1984).

Hartman, C. and Robinson, D. (2003). Evictions: The hidden housing problem. *Housing Policy Debate*, 14(4), pp. 461–501.

Helhedsplan Gellerup. (2018). Why we are doing it. [Online]. Available at: https://helhedsplangellerup.dk/english/why-we-re-doing-it/#1 [Accessed 15 November 2019].

Høghøj, M. (2019). *Between utopia and dystopia. A socio-cultural history of modernist mass housing in Denmark, c. 1945–1985.* PhD dissertation, Aarhus University, Aarhus.

Hooks, B. (2015). *Yearnings – Race, gender, and cultural politics.* Boston, MA: Routledge.

Jensen, L. (2013). Danmark – lokal boende demokrati och nationell korporatism. In: Bengtsson, B. (ed.) *Varför så olika? Nordisk bostadspolitik i jämförande historiskt ljus.* 2nd edn. Malmö: Égalité, pp. 49–117.

Jupp, E. (2017). Home space, gender and activism: The visible and the invisible in austere times. *Critical Social Policy*, 37(3), pp. 348–366.

Kallin, H. and Slater, T. (2014). Activating territorial stigma: Gentrifying marginality on Edinburgh's periphery. *Environment and Planning A*, 46(6), pp. 1351–1368.

Larsen, H. G. and Lund Hansen, A. (2015). Commodifying Danish housing commons. *Geografiska Annaler B*, 97(3), pp. 263–274.

Larsen, H. G., Lund Hansen, A., Macleod, G., and Slater, T. (2016). Introduction: The housing question revisited. *ACME*, 15(3), pp. 580–589.

Lees, L., Slater, T., and Wyly, E. (2008). *Gentrification*. London: Routledge.

LLO. (2008). Regeringsudvalg ønsker at svække beboerdemokratiet. Press release, the Tenants Organization of Denmark (*Lejernes Landsorganisation i Danmark*), Copenhagen 26 June.

Lund Hansen, A. and Clark, E. (2017). The right to freedom from displacement – Practical implementation and challenges. In: Davis, M. F., Gammeltoft-Hansen, T., and Hanna, E. (eds.) *Human rights cities and regions: Swedish and international perspectives*. Lund: Raoul Wallenberg Institute, pp. 23–30.

Lund Hansen, A., Larsen, H. G., Grydehøj, A., and Clark, E. (2015). Financialisation of the built environment in Stockholm and Copenhagen. *FESSUD Working Paper Series, No. 115* (ISSN 2052–8035), 82 pp.

Madsen, B. (2006). Skal den almene boligsektor gøres fri? *Samfundsøkonomen*, 4, pp. 31–34.

Mayer, M. (2016). Neoliberal urbanism and uprisings around Europe. In: Mayer, M., Thörn, C., and Thörn, H. (eds.) *Urban uprisings: Challenging the neoliberal city in Europe*. London: Palgrave Macmillan, pp. 24–41.

Ministry for Business and Growth. (2013). *Den finansielle krise i Danmark – årsager, konsekvenser og læring*. Copenhagen: The Government.

Ministry of Finance. (2004). *Fordeling og incitamenter 2004*. Copenhagen: Ministry of Finance.

Ministry of Housing, Urban and Rural Affairs. (2014). *Analyse – segregering i de fire største danske byomrad°er*. Copenhagen: Ministry of Housing, Urban and Rural Affairs.

Ministry of Social Affairs. (2006). *Den almene boligsektors fremtid. Rapport fra arbejdsgruppen vedrørende fremtidsperspektiver for en mere selvbærende almen sektor*. Copenhagen: Socialministeriet.

Molina, I. and Westin, S. (2012). *Renoviction – Even in Sweden. An effect of common benefit companies acting 'business-like'?* Paper presented at Association of American Geographers Annual Meeting, New York.

Møller, E. B. (2002). Almene boliger skal sælges. *Berlingske Tidende* (Daily edition), 2 January.

Piketty, T. (2014). *Capital in the twenty-first century* (Transl. Arthur Goldhammer). Harvard: Harvard University Press.

Porter, L. (2009). Planning displacement: The real legacy of major sporting events. *Planning Theory & Practice*, 10(3), pp. 395–418.

Rasmussen, L. L. (2018). New year television address (own translation). [Online]. Available at: https://www.bt.dk/politik/her-kan-du-laese-hele-lars-loekke-rasmussens-nytaarstale [Accessed 28 September 2018].

Schierup, C. U. (1994). The right to be different: Multiculturalism and the racialization of scandinavian welfare politics; the case of Denmark. *Innovation: The European Journal of Social Science Research*, 7(3), pp. 277–288.

Schneider, J. and Susser, I. (eds.) (2003). *Wounded cities: destruction and reconstruction in a globalized world*. Oxford: Berg.

Simonsen, K. and Koefoed, L. (2020). *Geographies of embodiment*. London: Sage.

Smith, D. (1994). *Geography and social justice*. Oxford: Blackwell.

Statistics Denmark. (2014). *Indkomster 2012. Tema: Indkomst og parforhold.* Copenhagen: Statistik Denmark.

The Government. (2018). *Et Danmark uden parallelsamfund – ingen gettoer i 2030.* [Online]. Available at: https://www.regeringen.dk/nyheder/ghettoudspil/ [Accessed 1 April 2019].

Wacquant, L. (2004). Ghetto. In: Smelser, N. J. and Baltes, P. B. (eds.) *International encyclopedia of the social and behavioural sciences.* Oxford: Elsevier, pp. 129–147.

Wacquant, L. (2008). *Urban outcasts: A comparative sociology of advanced marginality.* Cambridge: Polity Press.

Wacquant, L. (2015). Revisiting territories of relegation: Class, ethnicity and state in the making of advanced marginality. *Urban Studies*, 53(6), pp. 1077–1088.

Wacquant, L. (2016). *The two faces of the ghetto.* New York: Oxford University Press.

Wacquant, L., Slater, T., and Pereira, V. B. (2014). Territorial stigmatization in action. *Environment and Planning A*, 46(6), pp. 1270–1280.

9 Housing displacement in Australian cities

A Brisbane case study

Anne-Sophie Iotti

Introduction

Housing displacement is a growing problem internationally. It highlights the difficulty for low-income households to access affordable housing and tenure security, constrains their choices in location, and threatens a basic human right of decent housing for everyone. Brisbane is the capital of the state of Queensland and is located within the growing region of South-East Queensland. Moreover, it is the third largest city in Australia. This chapter explores displacement and its implication for low-income households through a case study of Brisbane, which has been the fastest growing city in Australia since the 1990s. Using a mixed methods approach, the study reports findings from investigating two triggers producing displacement in Brisbane: gentrification and urban renewal.

This chapter adopts the definition of displacement by Marcuse (1985: 205), based on Grier and Grier (1978).

> Displacement occurs when any household is forced to move from its residence by conditions that affect the dwelling or its immediate surroundings, and that: 1) are beyond the household's reasonable ability to control or prevent; 2) occur despite the household's having met all previously imposed conditions of occupancy; and 3) make continued occupancy by that household impossible, hazardous, or unaffordable.

Marcuse (1985) identifies four different forms of displacement: direct last resident displacement, direct chain displacement, displacement pressure, and exclusionary displacement. Displacement tends to be understood in its direct forms, but Marcuse's typology indicates that displacement can take indirect forms that can be equally devastating for the victims (Davidson, 2009). Therefore, this chapter will draw attention to direct as well as indirect types of displacement while investigating the case of Brisbane. To this end, the chapter's first section outlines the context of the case study and discusses the implications for displacement, and the second section highlights personal stories of displacement victims.

The transformation of Brisbane's inner city and its implication for displacement

Over the last fifty years, Australian cities have undergone significant changes related to population growth and development (Beer and Forster, 2002; Forster, 2006). Australian state capital cities in particular are undergoing urban renewal and gentrification, especially in the inner cities, which often contributes to the displacement of low-income households (Randolph, 2004; Atkinson and Easthope, 2009; Atkinson et al., 2011; Darchen and Ladouceur, 2013). Within the Australian context, studies have focused on Sydney and Melbourne, with little research investigating other state capital cities such as Brisbane, which for more than three decades has been one of the fastest growing metropolitan regions in Australia (Steele and Gleeson, 2010).

In the last thirty years, the city of Brisbane has undertaken many urban renewal projects that have led to gentrification in its inner city, but few precautions have been taken to avoid the displacement of low-income households (Darchen and Ladouceur, 2013). Two different processes led to the wider transformation of the city and of its inner city in particular. First, the old working-class housing stock went through a process of renovation, which is often associated with classical gentrification as defined by Glass (1964). Second, many urban renewal projects were taking place, largely driven by government (State of Queensland and Brisbane City Council) and private developers (Stimson et al., 2000; Walters and McCrea, 2014). These two processes are interconnected and often referred to as the 'inner city revitalization' (Stimson et al., 2000); importantly, the story of displacement characteristic of Brisbane unfolds with them.

According to Atkinson et al. (2011) and Weller and van Hulten (2012), displacement is also connected to wider housing pressures at the city scale, such as a lack of affordability and lack of security in the private rental market. Therefore, displacement is not only linked to gentrification in the Australian context. In many Australian cities, a shortage of affordable housing exists since housing prices have increased at a greater pace than average incomes, which pushes low-income households into housing stress (paying more than thirty per cent of their income on housing) (Atkinson et al., 2011; Weller and van Hulten, 2012). Housing affordability has declined in Brisbane; there is a shortage of affordable housing across the city, as well as difficulty to access the stock, especially for low-income households (Wulff et al., 2009; Hulse et al., 2014). For instance, Wulff et al.'s (2009) study reveals one affordable accommodation for every eight very low-income households. In addition, Liu (2011) reported that one third of the renters living in close proximity to the city centre are in housing stress. To provide a better picture, housing prices have increased on average 262 per cent since 1986 for Australia's eight capital cities, whereas average housing prices in Brisbane alone have increased by 299 per cent since 1986 (Geck and Mackay, 2018). Further, Geck and Mackay (2018) observed that housing price increases have occurred in two bursts – between 1987 and 1992 and, significantly, between 2001 and 2009. In particular, land values have grown under the pressure of gentrification and urban renewal. Thus, low-income households have been suffering from housing-cost pressures as well of shortage of affordable housing (Atkinson et al., 2011).

Moreover, Australia is characterized by a homeowner society (Ronald, 2008); homeownership is seen as a 'natural tenure', whereas renting is considered as 'a form of social deviance' (Atkinson and Jacobs, 2016: 3). Accordingly, renting is seen as inferior to owning. Therefore, the Australian housing market strongly favours homeowners, who benefit from the government's support. In contrast, the Australian private rental market does not provide secure occupancy for tenants, as it is considered a transitional phase prior to acceding homeownership (Hulse et al., 2011; Easthope, 2014). For instance, a lease commonly varies between six and twelve months; rental prices are based on housing-market prices and subject to increase every six months during the tenancy; and a lease can be terminated without grounds or on short notice (Hulse et al., 2011; Easthope, 2014). This lack of security in the private rental market results in renters moving frequently (Hulse et al., 2011). Within this context, private renters, especially those living on low-income, are at greater risk of displacement compared to other groups (Atkinson et al., 2011). The following subsections investigate displacement in the city of Brisbane after outlining gentrification and urban consolidation.

The early stages of the transformation of the inner city: renovation

According to Stimson et al. (2000), the first phase of the inner city's transformation began in the 1970s but really took off in the late 1980s. Brisbane's housing stock was characterized by single-detached housing on a plot of land, mostly made of timber and tin. This type of housing – commonly referred to as 'Queenslanders' – was characteristic of Brisbane's working-class housing. Often considered basic and cheap, it became target to the first phase of the transformation of the inner city, which was typified by the renovation of existing housing stock in working-class suburbs.

This renovation process has been driven by homeowners renovating the housing stock in working-class suburbs as well as local developers demolishing older stock and re-building it (Walters and McCrea, 2014). This has been accompanied by policies driven by the State of Queensland and Brisbane City Council to revitalize the inner city in order to increase its population and to accommodate population growth and employment opportunities based on 'urban boosterism' (Stimson et al., 2000). The population was declining in the inner city in the 1970s, contradicting the population growth of the South-East Queensland region (Stimson et al., 2000). Therefore, through these revitalization efforts, the inner-city population began to grow again by the late 1980s.

This growth was accompanied by population changes in the inner city. Stimson et al. (2000) studied changes in the population between 1986 and 1996 and observed a rise of the age group 20–54 years, with an even more pronounced rise of the sub-group 20–34 years. They also observed a decline in older age groups (such as 55–74 years) and in younger age groups (0–19 years). Additionally, they noted an increase in single-person households, as well as in couples without children. This went along with an increase of higher-income households in inner city locations. Finally, they reported a gain in renters and a decline in homeowners, which they associated with the growing number of students in the inner city. Thus, observable demographic changes show a gain of high-income households in the inner city to the detriment of low-income households.

In addition, Stimson et al. (2000) observed that housing prices have increased in the inner city, with average prices rising by almost 200 per cent from AUD 81,000 to approximately AUD 240,000 between 1992 and 1999. This housing price gain is due to increase in land value (Geck and Mackay, 2018). In short, the revitalization process in the inner city has led to changes in both the population and local housing prices, effectively resulting in a significant transformation. This process has been accompanied by urban renewal.

The second stage of the transformation of the inner city: urban consolidation

The second phase of the transformation of Brisbane's inner city is typified by urban renewal, which began in the early 1990s. Many urban renewals took place during this period, but here, urban consolidation will be used as the prime example. Importantly, urban consolidation is arguably a disguised form of state-led gentrification. This has already been shown by Bounds and Morris (2006) for the city of Sydney, where they connected gentrification with urban consolidation policies in the inner city. However, this chapter is not concerned by investigating the causes of displacement; rather, it is motivated by the fact that urban consolidation leads to displacement in Brisbane, which has not been explored yet.

Australian cities have been experiencing a process of suburbanization since the Second World War. This is strongly linked to the ideal of homeownership where the perfect house was a single detached house on a plot of land. This is often referred to as the 'Great Australian Dream' (Baum and Wulff, 2001). As a result, Australian cities are now typified by low density. This is even more pronounced in Brisbane, where the density (approximately 1,000 persons/sq km) is half as high as in other capital cities, such as Sydney (Raynor et al., 2017).

With regards to population growth, Brisbane's population has increased at a greater rate than in any other state capital, rising from 1.1 million inhabitants in 1986 to 2.3 million in 2004 (Cheshire et al., 2018), with population growth projections predicting up to 3.2 million by 2030 and 4.5 million in 2050 (Australian Bureau of Statistic, 2019). This rapid population increase has been driven by internal migration attracted by a 'sub-tropical urban lifestyle' (Stimson and Taylor, 1998; Clarke and Cheshire, 2018) and is tied to Brisbane being a 'sunbelt' metropolis (Guhathakurta and Stimson, 2007). Therefore, for the Queensland State government and Brisbane City Council, population growth is often framed as a severe issue that needs to be tackled (Raynor et al., 2017). Furthermore, this population growth has been mostly concentrated in outer city areas due to availability of affordable housing (Clarke and Cheshire, 2018). As a result, Brisbane has been growing together with neighbouring local governments along the two growth corridors, the Gold Coast and the Sunshine Coast (Stimson et al., 2000). This has led to concerns for the state government due to the unsustainability of low-density development: greenfield sites were converted into housing developments, which meant increased commuting time and car dependency for residents (Clarke and Cheshire, 2018). Therefore, governments were seeking a solution for how to accommodate this increasing population while also containing urban sprawl and delivering adequate infrastructures (Stimson et al., 2000; Atkinson and Easthope, 2009). Urban consolidation was seen as the solution to these problems.

Importantly, urban planning in Brisbane is driven by the state mandate on urban consolidation (Cheshire et al., 2018). In particular, 'the City Plan is formulated in accordance with the provisions set by the statutory South East Queensland Regional Plan 2009–2031, which guides growth and development in South-East Queensland and outlines compact development as a key regional priority' (Raynor et al., 2017: 1059). Therefore, local planning led by the Brisbane City Council applies state and regional policies (Raynor et al., 2017).

Urban consolidation is developed in all bands of the city (Coffee et al., 2016), but mostly in the inner city and along major transport corridors and activity centres (Queensland Department of Infrastructure and Planning, 2009), especially on brownfield sites and through infill redevelopments (Stimson et al., 2000; Cheshire et al., 2018). Principally, residential apartments and town houses are developed (Stimson et al., 2000). As a result, density has increased in the inner, middle, and outer city (Coffee et al., 2016). Nonetheless, differences exist across the city; additional high-rise dwellings have been mostly located in the inner city (Geck and Mackay, 2018), whereas the outer city is still characterized by low density (Coffee et al., 2016). In 2016, in the inner city, apartments represented 50.4 per cent and separate housing 41.4 per cent of dwellings (Australian Bureau of Statistic, 2016). This suggests that higher density dwellings are more concentrated in the inner city compared to the rest of Brisbane. In short, this phase has profoundly modified the physical fabric of the city, transforming it from low-density to high-density, especially in its inner city.

Both processes – renovation and urban consolidation – have contributed to changing the inner city of Brisbane. They have led to population increase in the inner city, thus reversing the previous trends of population decline. This population growth and, in particular, gain in professional households is often associated with 'internal housing market dynamics' rather than resulting from urban consolidation policy (Cheshire et al., 2018). Therefore, through the process of urban consolidation and renovation of the housing stock, the inner city has become attractive again and housing prices have increased (Stimson et al., 2000). Hence, the very same Brisbane that used to be known as Australia's largest country town is now likely to be referred to as 'Brisvegas' (Gleeson and Steele, 2010). In that sense, Brisbane is no longer considered a provincial and suburban city – a transformation that has been unique among Australian cities (Felton, 2011).

Inner city revitalization and implication for displacement

Inner-city revitalization is often portrayed as a success story. However, concerns arise over negative impacts. This subsection outlines issues with regard to displacement of low-income households. First, following the conversation on urban consolidation, Australian scholars pointed out the strong focus on attracting higher-income households without dependent children to live in inner cities (Stimson et al., 2000; Raynor et al., 2017). The target group is often identified as 'empty nesters', 'young professionals', and 'consumer-citizens' (Raynor et al., 2017).

Second, amenities have been improved in order to attract this specific demographic group. Museums, art galleries, and performing arts complexes, as well as restaurants and cafes – in short consumption services – have been developed in the inner city (Stimson et al., 2000). The marketing surrounding residential apartments in higher density emphasizes the benefits of living in central locations with easy access to amenities (Raynor et al., 2017). Effectively, a new lifestyle is constructed in the inner city, strongly targeting high-income groups (Stimson et al., 2000). This led Randolph (2006: 485) to warn about a potential 'urban spatial segregation based on lifestyle or life stage'.

Third, because developers aim to maximizing their profit, dwellings delivered in high density are likely to be small apartments of one or two bedrooms or luxurious apartments, thus beyond the financial means of many residents (Walters and McCrea, 2014). Therefore, it could be argued that apartments delivered through high density have an excluding effect on low-income households.

Fourth, Stimson et al. (2000) stressed in their study that growing housing prices in Brisbane had had a displacing effect on businesses and low-income households, especially for the elderly and welfare dependent. Moreover, Atkinson and Easthope (2009) showed the lack of answers to housing stress and the well-being of low-income households in Brisbane, while stressing that the focus is on high-income groups: 'It would seem that there is a benign ignorance of much of the social pain, loss and general discomfort that lower-income families have experienced during the long boom of the past decade' (Atkinson and Easthope, 2009: 75).

Finally, according to Raynor et al. (2017), government planners see displacement of low-income households as a result of urban consolidation as unavoidable. This suggests a normalization of displacement in the discourse of the professionals making the city. Therefore, concerns over displacement arise due to the type of dwellings built and the target population resulting from inner-city revitalization.

The stories of displaced households

This section explores stories of displacement tied with Brisbane's inner-city revitalization. It draws from the stories of White Australian private renters on low-income. It is important to stress again that gentrification and urban consolidation can occur simultaneously in the same neighbourhood.

Some interviewees explained how they feel displaced by neighbourhood changes while staying in place (i.e., displacement pressure). Gentrification and urban consolidation processes have modified the physical fabric of the neighbourhood – through renovations of the housing stock, demolition of existing housing (single detached houses), and replacement with new dwellings types, often high-rise apartments or townhouses. This has profoundly transformed the physical fabric of the neighbourhood by changing its character, which has impacted local residents as they feel disconnected with the new emerging aesthetic that stands in contrast with how they remember the place. Furthermore, some interviewees noticed that the social fabric of the neighbourhood was also changing, their social network disappearing, as a consequence. This change in population signals the arrival of high-income households – identified as coming from Sydney and Melbourne and being wealthier, White Australians, well-educated, and upper-middle class. For local residents, migrants from Sydney or Melbourne were symptomatic of the transformation of the suburb that they were witnessing, triggering a sense of loss of place.

The last neighbourhood change that the interviewees observed concerned commercial change. They noticed that local shops were closing and gradually being replaced by new shops (coffee shops, restaurants, and drinking venues, often franchise). Ordinary shops were supplanted by entertainment venues that also tended to be more expensive. More importantly, the interviewees sensed that these new shops were not designed for local residents but for outsiders and higher-income households. Few of the interviewees were visiting these new shops and often felt excluded. Additionally, the interviewees questioned the scale and pace of urban consolidation, noticing that Brisbane is going through rapid and important changes that profoundly modify the city. Some interviewees reported an intense process of redevelopment that seemed out of control and happening rapidly. Hence, the scale and rate of redevelopment appear to be sources of stress and anxiety for some of the interviewees.

These neighbourhood changes were coupled with economic pressures and lack of protection in the private rental market. For instance, the description of the case study has established that housing prices have massively increased in Brisbane. In an unregulated private-housing market, rents have increased in line with market prices. As a result, the interviewees reported that in most cases they had been priced out of their private dwellings due to constant rent increases (i.e., economic displacement). This was combined with a lack of protection in the private rental market, where increase in rents remains unchallenged and tenants are unable to seek protection. The constant rent increases were often coupled with a lack of maintenance of the rental property (i.e., physical displacement). It was difficult for renters to have repairs accomplished during their tenancy, or being afraid that the rent could increase again, they were reluctant to request any repairs to begin with. Nevertheless, the lack of maintenance coupled with constant increase in rent forced them out. Additionally, some private renters were evicted at the end of their leases as their landlords decided to sell their property to private developers for redevelopment purposes. As mentioned, landlords have the right to cease a lease at any time without any grounds. Therefore, the eviction could not be challenged by tenants on legal grounds. Nevertheless, it does result in their displacement.

These circumstances have triggered many private renters' displacement from their dwellings. Consequently, many of them tried to relocate within the same neighbourhood, but they faced several challenges while seeking a new accommodation – ranging from an overall rent-level rise at the neighbourhood scale or poor housing conditions, to property being sold for redevelopment purposes, to demolition of current housing – which resulted in an overall reduction of their options. However, the main challenge they faced was the reduced pool of affordable housing within the neighbourhood. Therefore, in order to remain, they were forced to accept higher mobility and tended to end up in poorer housing quality coupled with higher rent. As the pressure intensified, it was more and more difficult to remain. As a result, some were forced to relocate into more affordable nearby locations in an attempt to remain close to the previous neighbourhood (Atkinson et al., 2011), while some relocate to outer-city areas seeking affordability.

In short, economic pressures, lack of maintenance, shortage of affordable housing at the scale of the neighbourhood, lack of protection for tenants, and displacement pressures have forced out the interviewees. Indirect displacement was combined with direct displacement, which often took place prior to physical displacement. These results will be highlighted in line with an Australian study's findings regarding displacement.

Displacement in Australian cities

This section summarizes the main impacts of displacement and stresses the emotional cost associated with displacement. With regard to the impacts of displacement, Atkinson et al. (2011) and Weller and van Hulten (2012) observed that private renters on low-income households had been priced out of their gentrified neighbourhood. They showed that low-income households had found themselves in a very competitive housing market in which they suffer discrimination due to a preference toward young higher-income households over low-income households, despite being able to pay the rent. Importantly, these households stressed a decline in their housing quality to the extent of living in less secure forms of housing (overcrowded dwellings with family and friends, caravan parks, and so on).

The repercussion on people's lives was described by Atkinson et al. (2011) as the loss of social diversity, services, infrastructures, and social networks. They highlighted that those who had been displaced described a feeling of loss of place and a loss of friendships and networking. They argued that the loss of social networks had had a strong emotional impact on households that have been displaced. Therefore, while gentrification is often understood as a mere displacement of low-income households, a whole host of social and emotional costs exists.

Furthermore, regarding the emotional cost of urban consolidation, Cheshire et al. (2018) have shown that residents were strongly attached to low density settings. They also indicated that local residents were disrupted by the loss of character and heritage in their neighbourhood due to important physical changes. Finally, they observe several complaints by residents living surrounded by high-rise apartments, as their property is looked over by the high-rise apartments, in addition to the noise inherent to construction work (Clarke and Cheshire, 2018). The Brisbane City Council itself acknowledges that the increase of complaints about nuisances related to urban consolidation are caused by population growth and density increase, which in turn are a result of urban consolidation policies led by the State of Queensland (Clarke and Cheshire, 2018). Thus, urban consolidation evidently leads to feelings of discontent.

According to Atkinson (2015), neighbourhood changes such as demolition and re-building or the arrival of new shops and services are associated with the need for more affluent newcomers, not designed for local residents. Accordingly, these changes trigger a more distant relationship with the place, a feeling of 'subordination, discomfort and unease' (Atkinson, 2015: 10). Atkinson (2015) observed that households felt out of place, lost, and defamiliarized, even before they were forced displaced. The situation impacted them to a point of developing a feeling of un-homing. Atkinson (2015: 13) stressed the emotional cost attached to their displacement: 'The accounts of those who have been forced to move contain stories of loss, regret, and bitterness at the way in which their homes and the places they lived in have had to be abandoned'.

Conclusion

Australian capital cities are being redeveloped through urban renewal and gentrification, especially in their inner cities. However, the market-led urban consolidation too often contributes to the displacement of low-income households. This chapter has explained the main characteristics of displacement in Australian cities by investigating displacement in the city of Brisbane –which has never been studied, despite it being the fastest growing city in Australia – and the consequences of urban consolidation for local residents. The case study has attempted to show that urban consolidation associated with gentrification has profoundly transformed the city, which triggered feelings of discontent among local residents, as well as feelings of displacement for residents staying in place.

Within this context, the stories of displaced private renters reflect the process of displacement-while-in-place resulting from gentrification and urban consolidation; they also show the many pressures these renters faced through direct and indirect types of displacement. This was combined with a lack of protection from the private rental market, which made them more vulnerable to displacement pressures. Finally, while their displacement is due to gentrification pressures and urban consolidation, it is also due to housing pressures at the city scale – in particular, the lack of affordability. All these cumulative factors are playing out in their displacement.

References

Atkinson, R. (2015). Losing pne's place: Narratives of neighbourhood change, market injustice and symbolic displacement. *Housing, Theory and Society*, 32(4), pp. 1–18.

Atkinson, R. and Easthope, H. (2009). The consequences of the creative class: The pursuit of creativity strategies in Australia's cities. *International Journal of Urban and Regional Research*, 33(1), pp. 64–79.

Atkinson, R. and Jacobs, K. (2016). *House, Home and Society*. UK: Palgrave.

Atkinson, R., Wulff, M., Reynolds, M., and Spinney, A. (2011). Gentrification and displacement: The households impacts of neighbourhood change. *AHURI Final Report No.160*. Melbourne: Australian Housing and Urban Research Institute.

Australian Bureau of Statistic. (2016). *Census data* [Online]. Available at: https://quickstats.censusdata.abs.gov.au/census_services/getproduct/census/2016/quickstat/305?opendocument [Accessed 10 April 2019].

Australian Bureau of Statistic. (2019). *Population projections by region, 2017–2066: Population projections by region, Queensland* [Online]. Available at: http://stat.data.abs.gov.au/Index.aspx?Queryid=1080 [Accessed 28 June 2019].

Baum, S. and Wulff, M. (2001). Housing aspirations of Australian households. *AHURI Positioning Paper No.14*. Melbourne: Australian Housing and Urban research Institute.

Beer, A. and Forster, C. (2002). Global restructuring, the welfare state and urban programmes: Federal policies and inequality within Australian cities. *European Planning Studies*, 10(1), pp. 7–25.

Bounds, M. and Morris, A. (2006). Second wave gentrification in inner-city Sydney. *Cities*, 23(2), pp. 99–108.

Cheshire, L., Fitzgerald, R., and Liu, Y. (2018). Neighbourhood change and neighbour complaints: How gentrification and densification influence the prevalence of problems between neighbours. *Urban Studies*, 00(0), pp. 1–20.

Clarke, A. and Cheshire, L. (2018). The post-political state? The role of administrative reform in managing tensions between urban growth and liveability in Brisbane, Australia. *Urban Studies*, 55(16), pp. 3545–3562.

Coffee, N., Lange, J., and Baker, J. (2016). Visualising 30 years of population density change in Australia's major capital cities. *Australian Geographer*, 47(4), pp. 511–525.

Darchen, S. and Ladouceur, E. (2013). Social sustainability in urban regeneration practice: A case study of the Fortitude Valley Renewal Plan in Brisbane. *Australian Planner*, 50(4), pp. 340–350.

Davidson, M. (2009). Displacement, space and dwelling: Placing gentrification debate. *Ethics, Place and Environment*, 12(2), pp. 219–234.

Easthope, H. (2014). Making a rental property home. *Housing Studies*, 29(5), pp. 579–596.

Felton, E. (2011). Brisbane: Urban construction, suburban dreaming. *MC Journal: A Journal of Media and Culture*, 14(4), pp. 1–11.

Forster, C. (2006). The challenge of change: Australian cities and urban planning in the new millennium. *Geographical Research*, 44(2), pp. 173–182.

Geck, M. and Mackey, S. (2018). *Housing in Queensland: Affordability and preferences*. Queensland Productivity Commission.

Glass, R. (1964). *Introduction to London: Aspects of change*. London: Centre for Urban Studies.

Gleeson, B. and Steele, W. (2010). *A climate for growth: Planning South-East Queensland*. Brisbane: University of Queensland Press.

Guhathakurta, S. and Stimson, R. (2007). What is driving the growth of new 'Sunbelt' metropolises? Quality of life and urban regimes in Greater Phoenix and Brisbane-South East Queensland region. *International Planning Studies*, 12(2), pp. 129–152.

Hulse, K., Milligan, V., and Easthope, H. (2011). Secure occupancy in rental housing: Conceptual foundations and comparative perspectives. *AHURI Final Report No. 225*. Melbourne: Australian Housing and Urban Research Institute.

Hulse, K., Pawson, H., Reynolds, M., and Herath, S. (2014). Disadvantaged places in urban Australia: Analysing socio-economic diversity and housing market performance. *AHURI Final Report No. 225*. Melbourne: Australian Housing and Urban Research Institute.

Liu, Y. (2011). Spatial disparity in the private rental housing market in South East Queensland, Australia. *12th International Conference on Computers in Urban Planning and Urban Management*, CUPUM.

Marcuse, P. (1985). Gentrification, abandonment, and displacement: Connections, causes, and policy responses in New York City. *Journal of Urban and Contemporary Law*, 28, pp. 195–240.

Queensland Department of Infrastructure and Planning. (2009). *South East Queensland regional plan 2009–2031*. Queensland Government.

Randolph, B. (2004). The changing Australian city: New patterns, new policies and new research needs. *Urban Policy and Research*, 22(4), pp. 481–493.

Randolph, B. (2006). Delivering the compact city in Australia: Current trends and future implications. *Urban Policy and Research*, 24(4), pp. 473–490.

Raynor, K., Mayere, S., and Matthews, T. (2017). Do 'city shapers' really support urban consolidation? The case of Brisbane, Australia. *Urban Studies*, 55(5), pp. 1–20.

Ronald, R. (2008). *The ideology of home ownership: Home-owner societies and the role of housing*. New York: MacMillan.

Steele, W. and Gleeson, B. (2010). Planning in climate change. In: Gleeson, B. and Steele, W. (eds.) *A climate for growth: Planning South-East Queensland*. St. Lucia: UQP, pp. 123–135.

Stimson, R. and Taylor, S. (1998). Dynamics of Brisbane's inner city suburbs. *Australian Planner*, 35(4), pp. 205–214.

Stimson, R., Mullins, P., Baum, S., Davies, D., Gleeson, B., and Shaw, K. (2000). *Inner-cityrenaissance: The changing face, functions and structure of Brisbane's inner-city*. Brisbane: Techtrade.

Walters, P. and McCrea, R. (2014). Early gentrification and the public realm: A case study of West End in Brisbane, Australia. *Urban Studies*, 51(2), pp. 355–370.

Weller, S. and van Hulten, A. (2012). Gentrification and displacement: The effects of a housing crisis on Melbourne's low-income residents. *Urban Policy and Research*, 30(2), pp. 25–42.

Wulff, M., Dharmalingam, A., Reynolds, M., and Yates, J. (2009). Australia's private rental market: Changes (2001–2006) in the supply of, and demand for, low rent dwellings. *AHURI Positioning Paper No.122*. Melbourne: Australian Housing and Urban Research Institute.

10 Green gentrification and displacement in Barcelona

Melissa García-Lamarca, James Connolly, and Isabelle Anguelovski

Introduction

Increasing numbers of cities are planning and implementing urban green space in the face of environmental challenges, such as climate change, and to improve urban liveability from ecological, economic, and social standpoints. Research illustrating the multi-fold benefits of urban green spaces is extensive. Findings include reduced air pollution (Baró et al., 2014), increased biodiversity (Camps-Calvet et al., 2016), improved physical and mental health (Triguero-Mas et al., 2015; Gascon et al., 2016), and stronger local social ties (Connolly et al., 2013). From an economic development perspective, research has shown that abundant and high-quality green spaces strengthen the identity of an area as an attractive and desirable place to work and live and boost local economies and real estate values (Baycan and Nijkam, 2009; Dale and Newman, 2009; Anguelovski, 2015; Immergluck and Balan, 2018).

Nevertheless, a growing number of studies have found that the benefits urban green space brings are often unevenly distributed. As a result, the creation or restoration of green amenities in cities may not bring health and social benefits to *all* residents and, indeed, might lead to exclusion and displacement (Checker, 2011; Curran and Hamilton, 2012; Wolch et al., 2014; Goodling et al., 2015; Gould and Lewis, 2017). In the United States, where the majority of green gentrification research has taken place, researchers have found that wealthier and whiter neighbourhoods are those with the largest, best maintained, and most easily accessible green spaces (Wolch et al., 2005; Boone et al., 2009) or the most extensive urban forest canopy cover (Heynen et al., 2006). This is an issue of concern for equity and justice because such exclusionary dynamics, if not recognized and actively addressed, will likely only grow in the twenty-first century under the increasingly popular green and sustainable urban agenda (Connolly, 2019; García-Lamarca et al., 2019). In this light, urban greening might become a (new) prominent axis for generating deeper urban inequalities as it becomes embedded in urban capitalist development dynamics.

The pathway through which green amenities become drivers of displacement has been dubbed ecological, environmental, or green gentrification

(Dooling, 2009; Pearsall, 2010; Checker, 2011). This concept helps explain why and how improving the urban environment through greening measures can make historically disadvantaged residents vulnerable to displacement. As first defined by Sarah Dooling, environmental gentrification involves 'the implementation of an environmental planning agenda related to public green spaces that leads to the displacement or exclusion of the most economically vulnerable human population while espousing an environmental ethic' (Dooling, 2009: 630).

In this chapter, through a longitudinal and quantitative spatial analysis, we consider how green gentrification and concurrently displacement are manifesting in the city of Barcelona in the context of greening projects since the 1990s. The next section provides a more in-depth theoretical review of the literature on green gentrification and displacement, explaining how these topics are approached, while the subsequent section outlines Barcelona's greening trajectory since the democratic transition in the late 1970s. We then detail the spatial quantitative findings of our research in Barcelona to illustrate how housing and population trends have changed through time near eighteen new parks built in the working-class north-eastern part of the city since 1992. The key result to report here reveals a possible displacement of elderly and working-class residents from the more central-eastern areas of Barcelona toward northern, more affordable neighbourhoods. In the final section, we reflect on what the conceptual apparatus of green gentrification brings to studies of contemporary displacement.

Defining green gentrification and displacement

Gentrification is a process whereby low-income – and often people of colour – residents are displaced as a 'rent gap' is closed by investors achieving a rapid rise in cost of land and housing, thereby provoking sociocultural transformations of the neighbourhood (Clark, 1988; Smith, 1996). An emerging body of research examines how urban sustainability planning and processes of city re-naturing are incorporated into public-private redevelopment strategies that intensify gentrification processes. This research uncovers how green infrastructure serves as a catalyst for gentrification, and how the sustainability framework in which it is embedded both facilitates and conceals this process. The literature, at present largely situated in the United States and Canada, conceptualizes this phenomenon as green gentrification (Gould and Lewis, 2017), ecological gentrification (Dooling, 2009), and environmental gentrification (Pearsall, 2010; Checker, 2011; Curran and Hamilton, 2012; for some of the few cases to date outside of North America see Sandberg, 2014 and Anguelovski et al., 2019). Scholars in this area seek to understand how the benefits of greening are distributed. They often find that lower-income and non-White populations are systematically denied access to the benefits of urban greening, such as improved health and quality of life, as a consequence of larger development trends that lead to displacement from newly greened areas (Dooling, 2009; Anguelovski, 2014; Gould and Lewis, 2017; Anguelovski et al., 2019;).

Green gentrification literature, in its essence, highlights the socio-ecological underpinnings of processes of urban marginalization and exclusion. And while some city administrations are attempting to address inequities in greenspace access,[1] the creation of urban green amenities in low-income areas often generates a *green space paradox*. Neighbourhood-scale case studies so far have found that increases in the number, size, and quality of urban green spaces in areas facing gentrification pressures make these areas even more attractive and desirable, with both quantitative and qualitative research findings showing that greening contributes to the displacement of low income, people of colour, and otherwise disadvantaged populations (Dooling, 2009; Quastel, 2009; Pearsall, 2010; Checker, 2011; Curran and Hamilton, 2012; Goodling et al., 2015; Gould and Lewis, 2017). These populations then tend to occupy areas where green spaces are either scarce or poorly maintained (Heynen et al., 2006; Anguelovski, 2015). Meanwhile, real estate developers, public agencies, and local politicians tout green amenities as a competitive advantage, focusing on the environmental and economic benefits of greening and obviating the need for social and health benefits for all residents.

In the US context, several studies have shown how the remediation of former industrial sites, known as brownfields, makes a neighbourhood attractive for gentrification and causes displacement of the populations who suffered the consequences of industrial development, while richer homeowners and investors capture the gains from the appreciation of their property assets (Banzhaf and McCormick, 2006; Steil and Connolly, 2009; Essoka, 2010; Pearsall, 2010; Eckerd, 2011; Curran and Hamilton, 2012; Dillon, 2013; Anguelovski, 2014). While these studies have found that brownfield redevelopment can generate environmental gentrification, city-wide assessments of the impacts of environmental clean-up and associated green space creation are few in number. Most research conducted to date has considered indicators of *potential* displacement, like increased property values, rather than the actual displacement of residents (Pearsall, 2010; Immergluck and Balan, 2018).

Displacement has been found to take place in multiple ways in the green gentrification literature. Green space creation or rehabilitation leads to a 'Whitening' of neighbourhoods since African American and Latinx populations are displaced, as Gould and Lewis (2017) have found in various New York City boroughs. In Seattle, greening has directly led to the displacement of homeless people, whose notions of home and inhabitation of public green spaces do not coincide with ideological constructions of the same terms from state agencies and housed residents (Dooling, 2009). Patrick's (2014) queer ecological critique of New York City's iconic High Line uncovers the temporal and spatial displacement of queers and native plants that occurred through the High Line's development. The creation of Medellín, Colombia's forty-seven-square-kilometre greenbelt is generating a new landscape of pleasure and privilege for middle and upper classes as the livelihood territories of low-income, indigenous, self-built-housing residents are being grabbed for nature conservation and elite uses, leading to the physical, social, and ecological displacement of low-income indigenous residents (Anguelovski et al., 2019). Many of these cases illustrate how 'nature' is mobilized in an exclusionary fashion, following the pervasive notion of (White and well-to-do) conservationists who consider the environment as something pristine and unpeopled (Finney, 2014).

Several of the previously cited studies also point to a pernicious process of social, cultural, and procedural displacement that often underlies green gentrification. This manifests in many interconnected layers. One relates to how greening is desired (or not) since many people of colour may not feel included in a traditional park space (Wolch and Byrne, 2006) or might even see such spaces as sites of oppression and violence (Finney, 2014). Another is how (apparently) sustainable urban planning goals and processes displace the *possibility* of people of colour and low-income and disadvantaged populations to frame the broader political debate about justice, equity, and greening since participation in planning is limited to technocratic discussions of what a new green space should look like (Checker, 2011). Debates around what a green neighbourhood and city should be and who they are for are displaced by the idea that new green infrastructure projects are inherently good for all. This points to the critical importance of avoiding a post-political approach to urban sustainability (Swyngedouw, 2007). This means recognizing that greening is by no means a politically neutral goal but rather unfolds within profit-driven city-making dynamics and, like any urban process, is deeply connected to processes of exclusion, dispossession, and displacement of people of low income and people of colour.

Environmental gentrification, and the possible multifold displacement it entails, is an essential consideration for any urban sustainability model that seeks to promote both ecologically and socially responsible urban planning. Without this awareness, municipal representatives and sustainability advocates who uncritically accept calls for more urban green space may, against their own intentions, create new socio-spatial inequities (Pearsall and Pierce, 2010).

Now turning to our quantitative spatial analysis of Barcelona, is green gentrification unfolding in the city, and might it be leading to displacement? In the sections that follow, we take up this question through, first, providing a historical backdrop to greening in Barcelona and, then, highlighting the results of a spatial analysis of new parks created since 1990 across working-class districts of the city.

The transformation and greening of democratic Barcelona

In the 1970s, the legacy of the almost forty-year Franco dictatorship left many Spanish cities with a poor-quality built environment (Anguelovski, 2014). This included enormous deficits in housing, schools, cultural centres, health services, public transportation, and basic infrastructure (Saurí et al., 2009). While there were insufficient public parks and gardens, the most socially vulnerable areas of the city had a particularly acute lack of green space. Strong neighbourhood associations formed to protest for better public facilities and the improvement of neglected urban space in old and new urban peripheries (Calavita and Ferrer, 2004). After the first municipal democratic elections of 1979, Barcelona's City Council decided to prioritize increasing the number of parks and gardens through the implementation of new urban plans. Neighbourhoods were taken care of first as part of the overall strategy that sought 'the creation of a balanced and integrated Barcelona, without segregation, with social and territorial equality for all citizens', as stated by the city council (Casas cited in Calavita and Ferrer, 2004: 60). During this time, green spaces were primarily designed to provide meeting places for elderly residents and playgrounds for children (Saurí et al., 2009).

A new stage of urban redevelopment started when Barcelona won the bid for the 1992 Olympic Games (Anguelovski, 2014). Although the Olympics were initially seen as an opportunity to modernize Barcelona's infrastructure while retaining redistributive goals (McNeill, 1999), the scale of spatial redefinitions negated this approach. Instead, Barcelona's public green spaces shifted almost entirely toward the mega-event demands of the Olympics. The city council began to negotiate directly with the developers that were building the necessary infrastructure rather than with neighbourhood groups about the design and placement of green space. During this period of 'strategic urbanism', the social component present in the initial creation of public spaces during the early 1980s diminished (Monclús, 2003; Montaner, 2004; Anguelovski, 2014). Neighbourhood-scale social interactions became less of a priority seeing as new Olympic parks and facilities were designed more as aesthetic amenities for visitors than for residents. Consequently, the latter had fewer places for sitting and holding neighbourhood meetings or other social and cultural activities (Saurí et al., 2009).

Following the 1992 Olympic Games, the later stages of post-dictatorship redevelopment were characterized by efforts to leverage the newfound international appeal of Barcelona. During most of the 1990s, public park design and construction was often privately funded and was always tightly connected to economic development (Saurí et al., 2009; Montaner, 2004; Anguelovski, 2014). Toward the end of the 1990s, the city focused on redeveloping the remaining formerly industrial space in the Poble Nou neighbourhood (Sant Martí district). Near the newly built Universal Forum of Cultures event space, a luxury residential project was anchored by the second largest public park in Barcelona. Named the Diagonal Mar Park, it was marketed as the central component of the residential project's sustainability strategy, but it was widely criticized as being designed as an amenity for the high-end condominiums surrounding it (Anguelovski, 2014).

This redevelopment of Barcelona took place in the broader context of a housing boom, speculation, and 'mortgaged lives' for many working-class residents (García-Lamarca and Kaika, 2016). The housing boom reinforced the predominant tenure of homeownership in the city, although the Franco-era housing stock in many working-class districts meant that these areas were not as attractive to capital and thus not as prone to (green) gentrification dynamics as other more central parts of Barcelona with sizeable rent gaps. While widespread homeownership appeared to limit vulnerabilities to rental property-related evictions, which are significant due to very weak Spanish renter protection laws, the supposed security of mortgaged homeownership was a smokescreen as the profound 2007 real estate financial crisis provoked thousands of foreclosures and evictions across the city.

Understanding this evolution of urban greening in democratic Barcelona provides important context for us to then show how new environmental amenities in the city shape who benefits from urban life.

Green gentrification and displacement trends in Barcelona

In order to understand whether the distribution of new environmental amenities became equitable as Barcelona implemented its greening agenda, and whether this might have led to displacement, we examined how housing and population trends changed over time near eighteen new parks built in the northeast part of the city since 1992. We gathered data from various official city of Barcelona data sources at the smallest spatial unit available from 1992 to 2008[2] for: (1) home sale values, (2) household income, (3) percentage of population with a bachelor's degree or higher, (4) percentage of population over sixty-five years old living alone, (5) percentage of immigrant population whose nationality is from the Global North, and (6) percentage of immigrant population whose nationality is from the Global South. The latter was used as a proxy for non-White race and non-Caucasian ethnicity because neither race nor ethnicity data is gathered in Barcelona or Spain. The country as a whole has a relatively recent and intense history of immigration from countries in the Global South, which makes this variable indicative of people of colour in the city (Pelissier and Piñol, 2011). The main Global South immigrant groups in Barcelona arrived in the 1990s and early 2000s from Latin America (especially Ecuador, Bolivia, and Colombia), North Africa (Morocco, Algeria, Nigeria, and Senegal) and Asia (especially Pakistan, China, and the Philippines). Used in parallel with the percentage of immigrant population whose nationality is from the Global North – groups which tend to be Whiter, wealthier, and thus a more privileged population in terms of accessing housing – we were able to at least estimate possible racial and ethnic displacement related to urban greening.

Barcelona's high homeownership rate, explained previously, results in low residential mobility because people tend to stay in their homes for most of their lives. Vulnerabilities to displacement are thus greater for tenants and residents who can no longer afford their neighbourhood amenities and have difficulties going elsewhere to access basic services. Accordingly, key populations who are at greater risk of displacement in Barcelona include renters, elderly people living alone, working-class residents, and residents from the Global South.

For our spatial analysis, we mapped all new parks built in Barcelona between 1992 and 2004 (see Figure 10.1) and identified whether areas near parks experienced above average changes by comparing the trends within 100, 300, and 500 metre buffers of parks to those of the larger districts, where the parks are located over a four to twelve year time period (depending on when the park was built and the availability of data). Parks that are clustered together and built at the same time were merged into one buffer for the purpose of analysis, resulting in thirteen buffer areas reported in the results. We also verified the statistical significance of green gentrification trends by running global ordinary least squares (OLS) and local geographically weighted regressions (GWR) in order to confirm that parks were playing a causal role in the spatial variations we observe. The OLS regression model assumes spatial stationarity, that is, that the relationships between the sociodemographic variables and the distance to parks variable are equal across space. The GWR technique tested for variation across space. Since the city of Barcelona changed census tracts between years, models were

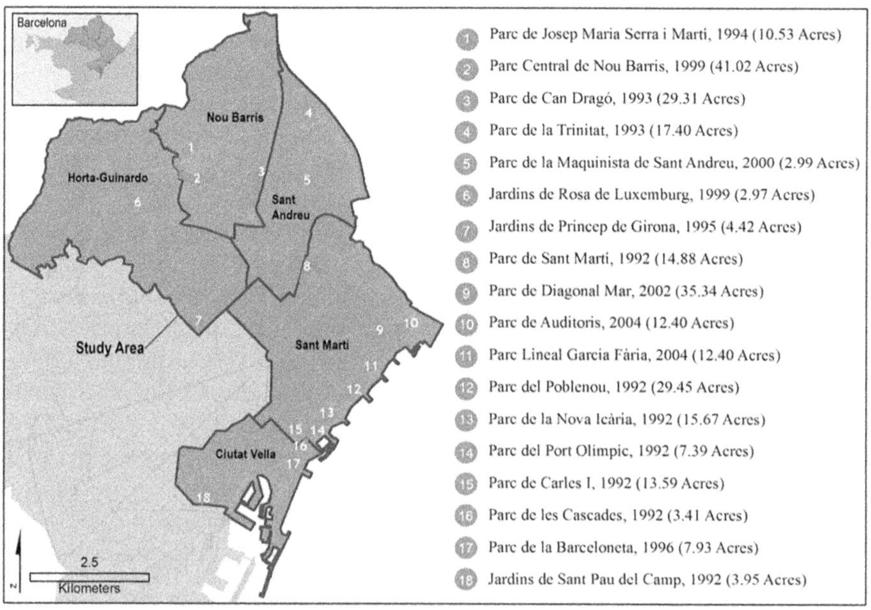

Figure 10.1 In green, parks built during the period 1992–2004 within the Ciutat Vella, Sant Martí, Sant Andreu, Nou Barris, and Horta-Guinardó districts.

Source: Anguelovski et al. (2018).

conducted for data from 2000 and from 2008, and regression coefficients were compared between the two years (see Anguelovski et al., 2018: 471–472 for further details). In the findings below, we present the general dynamics that these analyses demonstrated.

In order to determine the parks and gardens that appear to be associated with green gentrification, we allotted one point to parks with buffer areas that outpaced the districts where they are located in terms of the five sociodemographic variables studied, and then we added these points to form a composite score (Table 10.1). We identified areas near parks with greater increases than the district for the same period of time for the population with a bachelor's degree or higher and with the elderly population.

Table 10.1 Final green gentrification indicator scores for parks within the study area, with the eighteen parks merged to thirteen when parks were clustered together and built at the same time. Values in bold show areas where moderate (3) or strong (4) green gentrification appears to have occurred.

Park Name (Year Built)	*District*	*Bachelor's Degree*	*65+ Living Alone*	*Global North*	*Income*	*Home Sales*	*Total*
Jardins de Sant Pau del Camp (1992)	Ciutat Vella	0	0	0	1	0	1
Parc de la Barceloneta (1996)	Ciutat Vella	1	1	0	0	0	2
Jardins Princep de Girona (1995)	Horta-Guinardó	0	1	1	0	1	3
Jardins de Rosa de Luxemburg (1999)	Horta-Guinardó	1	0	0	0	0	1
Parc de Can Dragó (1993)	Nou Barris	1	0	0	0	1	2
Parc Josep M. Serra I Martí (1994)	Nou Barris	0	0	0	0	1	1
Parc de Nou Barris (1999)	Nou Barris	1	0	0	1	0	2
Parc de la Trinitat (1993)	Sant Andreu	1	0	0	1	1	**3**
Parc de la Maquinista (2000)	Sant Andreu	1	1	0	0	0	2
Parc de Sant Martí (1992)	Sant Martí	0	0	0	0	0	0
Parc del Poblenou (1992)	Sant Martí	1	1	1	1	0	**4**
Parc de Diagonal Mar (2002)	Sant Martí	1	1	0	1	0	**3**
Parc del Port Olímpic (1992)	Sant Martí	1	1	1	1	0	**4**

Source: Anguelovski et al. (2018).

For the percentage of immigrants from the Global North variable, we identified areas near parks with above average increases in Global North populations *and* below average increases in Global South populations. Income and home value variables are used interchangeably as the fourth variable due to data availability; we have one or the other for all parks, but not all data for all parks. Therefore, a score of four points would imply gentrification is occurring near parks across the full spectrum of indicators.

Using these indicators, we found that several parks in the Sant Martí district, where the former industrial Poble Nou neighbourhood is located, are associated with strong environmental gentrification (four out of four rating). These parks were built in a time of significant urban revitalization, driven by Barcelona's 1992 Olympic Games. In addition, other parks in the Sant Martí and Sant Andreu district built at different times – including the Diagonal Mar, Auditoris, and García Fària Lineal parks, and Parc de la Trinitat, respectively – experienced moderate environmental gentrification (rating three out of four). So did the Princep de Girona Garden in the southern area of the Horta-Guinardó district, located close to the desirable and gentrifying Gràcia district. The GWR findings support these results, showing that geography matters (Anguelovski et al., 2018). However, we did not find high green gentrification trends in all other parks located in the northwestern zone of Barcelona and in parts of the Ciutat Vella district (one or two out of four rating). Figure 10.2 below summarizes the results of our analysis.

A particularly interesting result that emerged concerns the elderly population. We found that the percentage of residents aged sixty-five and older living alone decreased closer to parks in both Ciutat Vella and Sant Martí districts. The strong spatial divide in the trend of these older residents echoes findings from other socio-economic variables showing a separate process of demographic change happening in costal parks compared to parks located further inland. Parks near the coast tend to show a greater decrease of elderly people living alone – as well as residents with no university degree, lower incomes, and born in the Global South – than parks in the larger districts. The general validity of the GWR model again shows that the spatial divide between more coastal and more inland parks is by no means a spatially random process. This result reveals that residents who are elderly, have low incomes, are without a bachelor's degree, and are from the Global South are possibly experiencing a displacement (levered by the creation of new green spaces) from the central-eastern areas of Barcelona toward the city's northern, more affordable neighbourhoods. Exploratory interviews with municipal staff similarly underlined that many residents from the city centre were displaced to the periphery during the 1990s and 2000s through processes of real estate expropriation or speculation. It seems from these results that parks function as another axis for the socio-demographic shifts we analysed.

While the northern areas of the city do have some green space, many are in close proximity to highways. These areas are generally less well connected to the city centre, and the overall housing and public equipment

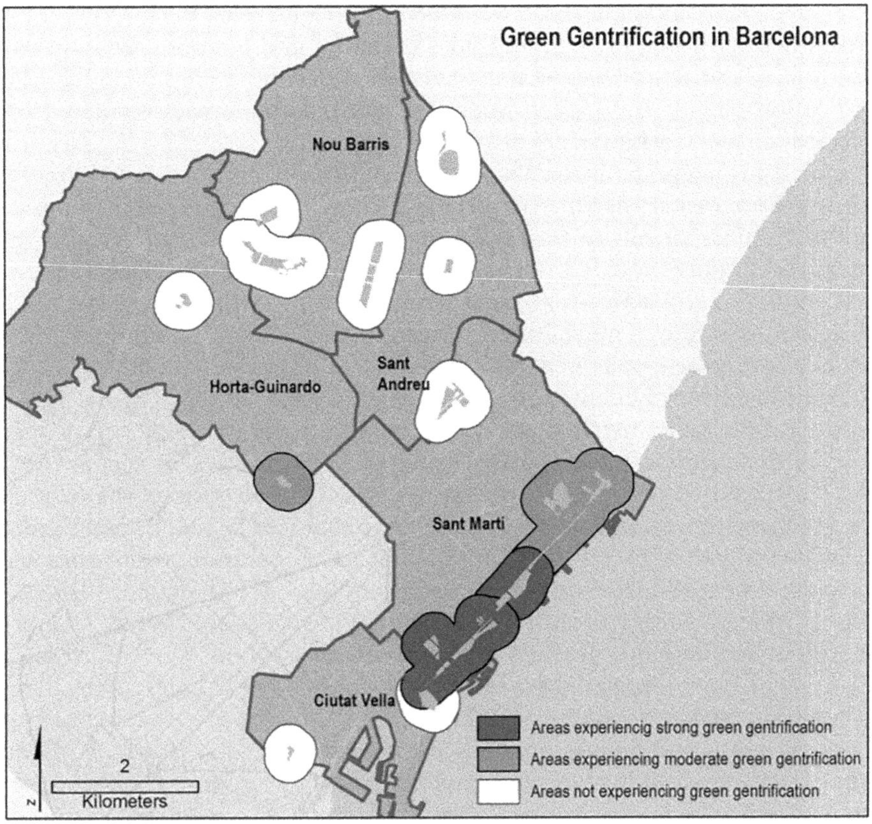

Figure 10.2 Areas where strong, moderate, and no green gentrification seem to be occurring.

Source: Anguelovski et al. (2018).

infrastructure are of lower quality than other parts of Barcelona. In these sites, residents might have good access to green space but not to other socio-environmental goods that are critical for broader well-being and urban liveability. Thus, Barcelona seems to illustrate a form of green goods polarization and re-segregation. Greener and more desirable neighbourhoods are populated by more privileged residents, while people of low income, people of colour, and elderly groups are displaced/confined into somewhat green but more socially fragmented and isolated neighbourhoods.

Conclusion

This chapter sought to introduce the paradoxical role that urban greening can play in cities, highlighting how attempts to physically green cities are not necessarily good for all as it can spur or exacerbate exclusion, marginalization, and displacement. We explained how the emerging literature on ecological, green, or environmental gentrification is exploring these dynamics. This literature mobilized our spatial and temporal quantitative analysis of the city of Barcelona, one of a handful of studies on green gentrification outside the United States, to identify the impacts of new park creation in key working-class districts.

Our study suggests that the impacts of park creation in socially vulnerable neighbourhoods depend on their context of creation, setting, and overall built environment. In Barcelona, it seems that green gentrification has occurred in parks located in more desirable neighbourhoods, such as the old industrial and waterfront areas within the Sant Martí and Ciutat Vella districts, or in the more central southern area of Horta-Guinardó. On the other hand, green gentrification appears to not have taken place in parks located in extremely dense, distressed neighbourhoods, such as the Raval in the Ciutat Vella district (which also tend to be much smaller parks), or in northern neighbourhoods with a semi-old building stock associated with the late dictatorship or early transition period. We also found that elderly people seem to be particularly vulnerable to physical displacement. Future studies could attempt to track displacement and population reconfiguration through an entire urban area, as population flows of low-income residents and people of colour from greened gentrifying neighbourhoods to more marginalized and also greened neighbourhoods (such as the northwestern districts of Barcelona) might create new forms of socio-spatial segregation. While this study focused on a longitudinal and quantitative spatial analysis of green gentrification and possible displacement processes, more in-depth qualitative research is required to understand if any potential cultural, social, or procedural displacement dynamics are intertwined in these green gentrification processes in Barcelona. Qualitative research is also required to explore how green amenities can be introduced to redeveloping districts like Sant Martí and Ciutat Vella without making them instruments for gentrification. Finally, considering the significance of tourism in the city of Barcelona, researching the impact of tourist rental flat locations in relation to green space and green gentrification dynamics is a pending task for future research.

Our findings and other green gentrification studies highlight how, despite the positive benefits brought by urban greening, new green spaces can also act as a (new) axis for urban exclusion because of the urban (re)development dynamics greening is embedded within. This stands in stark contrast with narratives touting greening as a solution for a more sustainable city, and it begs the question of how and for whom the city is being created. As a concept, green gentrification signals how green space can act as a catalyst for land and housing revalorization, displacing people of low income, people of colour, and other vulnerable populations.

Regarding deeper understandings of contemporary displacement processes, an important function of green gentrification as a critical conceptual apparatus is in uncovering the post-political nature of much sustainability and green planning taking place in cities around the world. This means that planning tends to take for granted the idea that 'all green is good' rather than questioning and opening up the broader planning and development framework to ask what type of greening neighbourhood residents, especially in working-class and minority neighbourhoods, need, and want. Consequently, the *possibility* for people of colour and low-income populations to frame the broader political debate about justice, equity, and greening is displaced because participation in planning is limited to technocratic discussions of what a new green space should look like. Thus, not only do residents face possible physical displacement, but they are also procedurally displaced beforehand through the planning process despite efforts to solicit resident input and participation in planning.

As awareness about green gentrification is growing. Some communities are exploring a 'just green enough' strategy (Curran and Hamilton, 2018), where gentrifying actors and long-term residents work together to remediate and redevelop contaminated sites without displacement. Such strategies are in their infancy and move against solely profit-driven development dynamics; therefore, they require community organization and mobilization. However, 'just green enough' strategies might be a temporary stop gap, highlighting the need for housing-centred strategies to combat displacement. Furthermore, the leading examples of 'just green enough' approaches also require a large infrastructure of polices to combat displacement when greening occurs, which generally takes decades of organizing. While such bottom-up resistance strategies are certainly crucial, this chapter points to the importance of integrating a critical vision of urban greening into planning and development processes to mitigate the potential exclusionary effects of making a physically greener city.

Notes

1 For examples, see the Greening without Gentrification: Learning from Parks-Related Anti-Displacement Strategies Nationwide policy report by Rigolon and Christensen. https://www.ioes.ucla.edu/project/prads/.

2 The 2008 real estate financial crisis resulted in a sigificant drop in housing prices and substantial emigration from the city. We therefore stopped our analysis at this date.

References

Anguelovski, I. (2014). *Neighborhood as refuge community reconstruction, place remaking, and environmental justice in the city*. Cambridge, MA: The MIT Press.

Anguelovski, I. (2015). From toxic sites as LULUs to green amenities as LULUs? New challenges of inequity, privilege, and exclusion in urban environmental justice. *Journal of Planning Literature*, 31(1), pp. 23–36.

Anguelovski, I., Connolly, J., Masip, L., and Pearsall, H. (2018). Assessing green gentrification in historically disenfranchised neighborhoods: A longitudinal and spatial analysis of Barcelona, Spain. *Urban Geography*, 39(3), pp. 458–491.

Anguelovski, I., Irazábal-Zurita, C., and Connolly, J.J.T. (2019). Grabbed urban landscapes: Socio-spatial tensions in green infrastructure planning in Medellín. *International Journal of Urban and Regional Research*, 43(1), pp. 133–156.

Banzhaf, H. and McCormick, E. (2006). Moving beyond cleanup: Identifying the crucibles of environmental gentrification. *Andrew Young School of Policy Studies Research Paper Series No. 07–29*. Atlanta: Georgia State University.

Baró, F., Chaparro, L., Gómez-Baggethun, E., Langemeyer, J., Nowak, D. J., and Terradas, J. (2014). Contribution of ecosystem services to air quality and climate change mitigation policies: The case of urban forests in Barcelona, Spain. *Ambio*, 43(4), pp. 466–479.

Baycan-Levent, T. and Nijkamp, P. (2009). Planning and management of urban green spaces in Europe: Comparative analysis. *Journal of Urban Planning and Development*, 135(1), pp. 1–12.

Boone, C., Buckley, G., Grove, M., and Sister, C. (2009). Parks and people: An environmental justice inquiry in Baltimore, Maryland. *Annals of the Association of American Geographers*, 99(4), pp. 767–787.

Calavita, N. and Ferrer, A., (2004). Behind Barcelona's success story – Citizen movements and planners' power. In: Marshall, T. (ed.) *Transforming Barcelona: The renewal of a European Metropolis*. London: Routledge, pp. 48–66.

Camps-Calvet, M., Langemeyer, J., Calvet-Mir, L., and Gómez-Baggethun, E. (2016). Ecosystem services provided by urban gardens in Barcelona, Spain: Insights for policy and planning. *Environmental Science and Policy*, 62, pp. 14–23.

Checker, M. (2011). Wiped out by the 'greenwave': Environmental gentrification and the paradoxical politics of urban sustainability. *City and Society*, 23(2), pp. 210–229.

Clark, E. (1988). The rent gap and transformation of the built environment: Case studies in Malmö 1860–1985. *Geografiska Annaler. Series B. Human Geography*, 70(2), pp. 241–254.

Connolly, J. J. T. (2019). From Jacobs to the Just City: A foundation for challenging the green planning orthodoxy. *Cities*, 91, pp. 64–70.

Connolly, J. J., Svendsen, E. S., Fisher, D. R., and Campbell, L. K. (2013). Organizing urban ecosystem services through environmental stewardship governance in New York City. *Landscape and Urban Planning*, 109(1), pp. 76–84.

Curran, W. and Hamilton, T. (2012). Just green enough: Contesting environmental gentrification in Greenpoint, Brooklyn. *Local Environment*, 17(9), pp. 1027–1042.

Curran, W. and Hamilton, T. (eds.) (2018). *Just green enough: Urban development and environmental gentrification*. Abingdon and New York: Routledge.

Dale, A. and Newman, L. L. (2009). Sustainable development for some: Green urban development affordability. *Local Environment*, 14(7), pp. 669–681.

Dillon, L. (2013). Race, waste, and space: Brownfield redevelopment and environmental justice at the Hunters Point Shipyard. *Antipode*, 46(5), pp. 1205–1221.

Dooling, S. (2009). Ecological gentrification: A research agenda exploring justice in the city. *International Journal of Urban and Regional Research*, 33(3), pp. 621–639.

Eckerd, A. (2011). Cleaning up without clearing out? A spatial assessment of environmental gentrification. *Urban Affairs Review*, 47(1), pp. 31–59.

Essoka, J. D. (2010). The gentrifying effects of brownfields redevelopment. *Western Journal of Black Studies*, 34(3), pp. 299–315.

Finney, C. (2014). *Black faces, White spaces: Reimagining the relationship of African Americans to the great outdoors*. Chapel Hill: University of North Carolina Press.

García-Lamarca, M., Anguelovski, I., Cole, H., Connolly, J., Argüelles, L., Baró, F., Loveless, S., Pérez del Pulgar, C. and Shokry, G. (2019). Urban green boosterism and city affordability: For whom is the 'branded' green city? *Urban Studies*. doi:10.1177/0042098019885330.

García-Lamarca, M. and Kaika, M. (2016). 'Mortgaged lives': The biopolitics of debt and housing financialisation. *Transactions of the Institute of British Geographers*, 41(3), pp. 313–327.

Gascon, M., Triguero-Mas, M., Martínez, D., Dadvand, P., Rojas-Rueda, D., Plasència, A., and Nieuwenhuijsen, M. J. (2016). Residential green spaces and mortality: A systematic review. *Environment International*, 86, pp. 60–67.

Goodling, E., Green, J., and McClintock, N. (2015). Uneven development of the sustainable city: Shifting capital in Portland, Oregon. *Urban Geography*, 36(4), pp. 504–527.

Gould, K. A. and Lewis, T. L. (2017). *Green gentrification: Urban sustainability and the struggle for environmental justice*. Abingdon and New York: Routledge.

Heynen, N., Perkins, H. A., and Roy, P. (2006). The political ecology of uneven urban green space. *Urban Affairs Review*, 42(1), pp. 3–25.

Immergluck, D. and Balan, T. (2018). Sustainable for whom? Green urban development, environmental gentrification, and the Atlanta Beltline. *Urban Geography*, 39(4), pp. 546–562.

McNeill, D. (1999). *Urban change and the European Left: Tales from the new Barcelona*. London: Routledge.

Monclús, F. J. (2003). The Barcelona model: And an original formula? From 'reconstruction' to strategic urban projects (1979–2004). *Planning Perspectives*, 18(4), pp. 399–421.

Montaner, J. Ma. (2004). La evolución del modelo Barcelona (1979–2002). In: Borja, J., Muxí, Z., and Cenicacelaya, J. (eds.) *Urbanismo en el siglo XXI: una visión crítica: Bilbao, Madrid, Valencia, Barcelona*. Barcelona: Escola tècnica superior d'arquitectura de Barcelona Edicions UPC.

Patrick, D. J. (2014). The matter of displacement: A queer urban ecology of New York City's High Line. *Social and Cultural Geography*, 15(8), pp. 920–941.

Pearsall, H. (2010). From brown to green? Assessing social vulnerability to environmental gentrification in New York City. *Environment and Planning C: Government and Policy*, 28(5), pp. 872–886.Pearsall, H. and Pierce, J. (2010). Urban sustainability and environmental justice: Evaluating the linkages in public planning/policy discourse. *Local Environment*, 15(6), pp. 569–580.

Pelissier, F. and Piñol, T. (2011). *Geografia de l'Arrelament Social en el Municipi de Barcelona, Un Estat de la Qüestió*. Barcelona: Fundació Adsar.

Quastel, N. (2009). Political ecologies of gentrification. *Urban Geography*, 30(7), pp. 694–725.

Sandberg, L. A. (2014). Environmental gentrification in a post-industrial landscape: The case of the Limhamn quarry, Malmö, Sweden. *Local Environment*, 19(10), pp. 1068–1085.

Saurí, D., Parés, M., and Domene, E. (2009). Changing conceptions of sustainability in Barcelona's public parks. *Geographical Review*, 99(I), pp. 23–36.

Smith, N. (1996). *The new urban frontier: Gentrification and the revanchist city.* London and New York: Routledge.
Steil, J. and Connolly, J. (2009). Can the Just City be built from below? Brownfields, planning, and power in the South Bronx. In: Marcuse, P., Connolly, J., and Novy, J. (eds.) *Searching for the just city: Debates in urban theory and practice.* London: Routledge, pp. 173–193.
Swyngedouw, E. (2007). Impossible 'sustainability' and the postpolitical condition. In: Krueger, R. and Gibbs, D. (eds.) *The sustainable development paradox: Urban political economy in the United States and Europe.* New York: Guilford Press, pp. 13–40.
Triguero-Mas, M., Dadvand, P., Cirach, M., Martínez, D., Medina, A., Mompart, A., and Nieuwenhuijsen, M. J. (2015). Natural outdoor environments and mental and physical health: Relationships and mechanisms. *Environment International,* 77, pp. 35–41.
Wolch, J. and Byrne, J. (2006). Nature, race, and parks: Past research and future directions for geographic research. *Progress in Human Geography,* 33(6), pp. 743–765.
Wolch, J., Byrne, J., and Newell, J. P. (2014). Urban green space, public health, and environmental justice: The challenge of making cities 'just green enough'. *Landscape and Urban Planning,* 125, pp. 234–244.
Wolch, J., Wilson, J. P., and Fehrenbach, J. (2005). Parks and park funding in Los Angeles: An equity-mapping analysis. *Urban Geography,* 26(1), pp. 4–35.

11 A dialogue on displacement as entwined colonialism

Matt Hern and Lisa K. Bates

Introduction

This chapter is a conversation between two people who are interested in gentrification, displacement, and dispossessions from land. We come from very different backgrounds and perspectives but share friendship, critical sensibilities, and research interests.

Lisa is a Black American academic in urban planning at Portland State University. She is involved with community anti-displacement organizing in the city of Portland. Matt is a middle-aged white-dude community organizer who founded Solid State Community Industries in Surrey, British Columbia (Canada).

We met some years ago in Portland and have had an ongoing series of conversations in various forms. There are many significant confluences and complementary tendencies in our thinking, and also some intriguing points of friction. This chapter is part of a larger back-and-forth around racial capitalisms, tenure, property, and land allocations. In recent years, there has been an ongoing global surge in activist and theoretical interest in displacements, expulsions, and dispossessions from land, but to our minds, there are some critical gaps and tensions in most theoretical explorations of displacement that are worthy of further exploration. We instigated this conversation to start to think through some of our most intransigent theoretical blockages, to look for help in thinking through tough issues, and to enjoy each other's company and ideas.

LKB

Starting the conversation about gentrification with a *longue durée* approach to the history of dispossession and racialization in North America is essential to thinking through how we make more just cities. I will perhaps depart from some of the prominent Indigenous theorists you often rely on, Matt, in saying that Indigenous *and* Black struggles must be placed at the fore and understood as deeply intertwined by the history of settler colonialism. North American wealth was built in the settler colonial project by the violent theft of both land and labour – the labour being enslaved Africans, removed from their/our own Indigenous lands that were being exploited for resources. That transit across the Atlantic rendered Black people as chattel – nonhuman property. We were not savages to be tamed; we were the plow horse – or even the plow. Indigenous African culture ways (multiple and distinct) were severed almost completely as part of this total dehumanization.

Black labour on North American Indigenous land built not only wealth but a new status of citizenship for white settlers in the fledgling democracy of the United States. The New Orleans art collective Blights Out draws the connection from the slave auction to the land auction, reminding us that the transformation from human to chattel was in service of white political claims:

> If the buying and selling of property is a qualification of citizenship, then the auction of real estate is the process by which citizenship and the freedoms and limitations thereof are established, maintained and revoked. Through the dehumanization of human-turned-property, a U.S. citizen with rights and 'freedoms' is born. This dehumanization is normalized through the spectacle of the auction.
>
> (Kinney, 2017)

In the United States, this status of chattel transferred inter-generationally, and even after 'freedom', descendants of enslaved Africans were never to be eligible for citizenship (the essential finding of the US Supreme Court's 1857 *Dred Scott* case decision). Once a body as property, the Black person could never become part of the body politic; the long tail of this foundational principle is seen in the attenuated citizenship of Black people today, as we see through the continued struggle for voting rights and to end police violence, among other movements.

With this link between ownership and citizenship, as well as the fundamental issue of self-sufficiency and self-reliance outside of labour in service to whites, Black claims to land began immediately post-Emancipation from slavery after the US Civil War. The broken promise of 'forty acres and a mule' resonates today in claims to reparations that emerge from the long history of denial of wealth generation through homeownership (e.g., from Ta-Nehisi Coates's widely circulated 2014 article 'The Case for Reparations'). In the United States (and maybe Canada?), property ownership continues to convey tremendous financial and political advantages, and it has been seen as fundamental not only to self-reliance but also, fundamentally, to safety in a white supremacist society. Blacks have long sought to create their own spaces, build their own towns, and either individually or collectively own land as a refuge – even when their 'ownership' was not recognized by the 'official'/racist institutions of registry. These spaces – neighbourhoods, towns, or cooperatively owned farms – have served as places of memory keeping within a dominant culture that has sought to fragment family and community histories as a dangerous source of power. Scholar Andrea Roberts (2018) writes about 'Freedom Colonies' in Texas, often physically hidden enclaves of homes, where Founders Day celebrations involve recounting the origin story, almost invariably of escaped slaves finding a secluded space to settle. She describes how this performance of the cultural repertoire is critical to reproducing identity as Black landholders, as autonomous people, and to making people visible in place, on their own terms (Roberts, 2018).

Of course, even owning property has not historically given Black people in North America a place of refuge, in any context. Black towns like Eatonville and Rosewood were burned by white supremacists who could not countenance Black self-sufficiency. Rural African Americans were discriminated against by the US Department of Agriculture and lost farm land, including the first community land trust, the New Communities Initiative, which cooperatively owned and farmed 6,000 acres in Georgia from 1969 to its foreclosure in 1985 (Douglas, 2017). Roberts (2018) documents how the mainstream historical preservation process denigrates freedom colonies' knowledge of their origins and the possibility of unbroken oral history as a legitimate source for documenting meaning, refusing historic status that could preserve these towns. Most pertinent to the gentrification discussion, city after city has pursued urban renewal practices that destroyed Black neighbourhoods physically with infrastructure and economic development projects, displacing and dispossessing Black people of even the under-served areas they were able to claim.

Thus, from some perspectives for Black politics, property ownership has been seen as vitally important to citizenship claims and reparations; at the same time, it has never been available on equal footing. Attenuated forms of property ownership have even been damaging for Black stability, as in the massive losses during the foreclosure crisis in the United States. Place has been deeply meaningful to identity and autonomy, but it has also been under continual external attack – both an attack on what that place/identity meaning is, through the denigration of Black spaces as ghetto and savage, and literal attacks on the existence of Black spaces as potentially self-determining, through economic and physical violence and the violence of state planning.

In their 2007 book *Black Geographies and a Politics of Place*, Black geographers Katherine McKittrick (Canadian) and Clyde Woods (American) describe a fundamental challenge for Black people in North America as living 'in and through social systems that reward us for consuming, claiming, and owning things' (McKittrick and Woods, 2007: 5). For Black North Americans, the challenge is to grapple with the reality that property ownership has not brought full citizenship to fruition, to confront the colonial project more explicitly, and to reconceptualize property itself. Ownership of slaves and ownership of land were obviously intertwined since Black labour made possible the development of land expropriated from Indigenous people.

To quote the Black Land and Liberation Initiative, part of the US-based Movement for Black Lives,

> We continue the legacy of those who have fought before us and have sought the return of stolen accumulated wealth ... The basis of all wealth is the combination of land and labor, and to be self-determining we must liberate both.
>
> (Tutashinda, 2017)

How can we learn across and through the Black and Indigenous experiences as twinned under settler colonialism? It seems apparent to me that de/colonization on this continent requires grappling with the enslavement of people, and with specific and particular attention to the condition of Afro-descended people. However, the two conversations seem rarely to engage with one another – and I feel a visceral pain when I attend events that open with an acknowledgement of Indigenous lands and people without recognizing chattel slavery and its legacies. Am I just being egocentric in my feelings? Or is this a productive instinct towards how we might begin to remake our cities in an era of expropriation, gentrification, and displacement? How can we learn across borders – incorporating the Canadian focus on Indigenous people and the US focus on racialized Black folks?

MH

Hey, thanks for all this, LKB. I think it is compelling, and there is so much brilliance here. I think I want to start by picking up on the property theme a little. Maybe it is a thread worth pulling on in partial answer to your last question. I am not sure about the national bifurcation you imply here, but I think the heart of the question is critical.

I am curious about the location of property theory – Lockean and otherwise – as a rationality that cannot see outside of itself and views everything around it as subject to owner/owned gazes: land, bodies, animals, air, labour, whatever, or whoever. Once property is substantiated as an operational way of being in the world, then everything is just another market. Property is by definition dominatory. It is not, and cannot ever be, a decent frame to view land, humans, or the other-than-human. Any recourse to property, explicit or presumed, is to revert to narratives of ownership, possession, and control. Rob Nichols (as usual) has articulated this issue brilliantly in his 2018 essay 'Theft is Property: The recursive logic of dispossession'. He writes,

> I focus in particular on the persistent concern that insofar as dispossession and expropriation gain their normative force from a perceived violation or corruption of actually existing property relations (i.e., a species of theft), they are generally conservative concepts that moreover

tend to reinforce a proprietary model of social relations that critical theorists generally seek to undermine.

Judith Butler once questioned it as follows: 'Is possession the answer to dispossession?' (Butler and Athanasiou, 2013). The answer has to be 'not at all'. As Nichols (2018) highlights, the frame of dis-possession presumes the wrongful disruption of a previously existing rightful possession, which then fixes particular kinds of social relations in place, not only as desirable but as 'natural'.

(Although this essay is good in so many ways, as you point out, it is totally baffling how an essay that has a huge section on Alabama and dispossession of Indigenous peoples fails to substantively mention Black people at all. This is extremely weird, especially for someone who I respect as much as Rob, and it points to the critical importance of your points above, specifically the last question.)

In particular, the contemporary North American fixation on homeownership is a manufactured phenomenon, one that is of recent (twentieth century) provenance and that has been specifically designed to augment and exacerbate racial and class divisions. As Desmond (2017: 5) has put it, speaking of the United States, 'The owner-renter divide is as salient as any other in this nation, and this divide is a historical result of statecraft designed to protect and promote inequality'.

I can see why so many people are desperate to own: renting on the private market is a grim proposition. As a renter my whole life, I know (and am reminded at the outset of every month) that I actually *am* buying property, just for someone else. I experience the precarity, the hole you throw your money into, the lack of any security of tenure, the occasional harassment, the anxiety, et cetera. Nevertheless, possession is the wrong response to a precarity of tenure.

When I give talks, it is very common for someone (very often a young Black person) to comment that homeownership is what saved their family, gave them a rock (physically and socially) to rely on, and that ownership is the one trustworthy constant that has allowed them social mobility. This is true for so many people: homeownership is perhaps the best way to accrue equity via enforced savings and occasional land-lifts, and it does provide particular kinds of anxiety-reducing certainty. There is no disputing this. However, on a social scale, the dream of a nation of homeowners is a nightmare of deepening cycles of inequality, commodification of every nook and cranny of our lives, and the relentless fetishization of property. Regimes of homeownership provide lifeboats for some, but only while raising the tide of inequality.

Thus, what do we say to those who have and continue to claim that property is the only real route to economic self-determination and independence? Part of my response is that the *rent vs own* bifurcation is a mirage, a shell game of false choices. Security of tenure can be achieved in all kinds of ways, and imaginative policy wedded to aggressive non-market politics can achieve the security and independence that ownership claims while disavowing profiteering from land.

Not just in an everyday sense but ontologically as well, the fixation on *ownership* is terrible for us all.

Starting this argument by foregrounding originary land and body thefts breaks the mirage for many of us. As Michi Saagiig Nishnabeg scholar Leanne Simpson (2013) puts it, 'Under Canadian law, if you steal a car – even if you really like it and its really nice – if you get caught, you still have to give it back'. That is just true and squaring up to that means disrupting all our notions of land and property. To me, that claim seems so hopeful.

However, I would agree that it is impossible to speak of land reparations without simultaneously speaking of reparations for slavery and Black genocides. These huge blocks of power and money all around us are resting on the twin thefts of land and bodies, and it is not obscure or undocumented or vague what that theft constituted. The phrase 'The basis of all wealth is the combination of land and labor, and to be self-determining we must liberate both' is beautifully put, but I guess our question here is, what does 'liberate' mean.

With regards to the question you ask, LKB, 'How can we learn across borders …', it strikes me that it implies two separate kinds of battles that need to support one another. Would it be too much to suggest that they are actually inseparable – entangled via rationalities of property, each reinforcing the other?

LKB

It is not too much, and it strikes me as utterly necessary. The public dialogue around the concept of reparations for Black American descendants of slaves has ramped up ('ADOS' is our new catchy nickname). It is even appearing in 2020 presidential candidate platforms! The typical invocation is to 'forty acres and a mule' – but it has not been interrogated in terms of the relationship of that never-realized promise to the land itself in the settler colonial nation-state. Not only do you not get to keep the stolen car, but you do not get to exonerate yourself by giving it away to somebody else, even if that person really, really needs a ride. A reparation based on being 'repaid' in Indigenous land does not undo the fundamental dehumanization. As Tuck and Yang (2012: 30) point out, this version of reparation is really about including Black folks into the settler identity (even if only partially and still as a racialized, minoritized 'lesser'):

> [I]t is no accident that the U.S. government promised 40 acres of Indian land as reparations for plantation slavery … we urge you to consider how enslavement is a twofold procedure: removal from land and the creation of property (land and bodies). Thus, abolition is likewise twofold,

requiring the repatriation of land and the abolition of property (land and bodies). Abolition means self-possession but not object-possession, repatriation but not reparation.

MH

Just recently, I met with some friends from the Hogan's Alley Working Group who are organizing an exciting new housing development – and I think the story is highly germane to our conversation.

Hogan's Alley was the centre of Vancouver's Black community from the early twentieth century. It coalesced on the edge of Strathcona, due to its proximity to the rail yards where many Black labourers could find employment, and it promised some respite from the hostility and discrimination Blacks faced from residents in so many other neighbourhoods. The district was always under threat and harassment from authorities for the breadth of its existence due to the perceived (and some real) illegalities that concentrated there, but the area was the beating heart of Vancouver's Black community – with restaurants, clubs, services, businesses, churches, and a vibrant social scene. It has always been incredibly difficult for Black people in Vancouver, and it is important not to be nostalgic about some imagined halcyon past when the Black community was secure and welcomed. Nevertheless, Hogan's Alley was Vancouver's first (and still only) clearly Black neighbourhood. It was slowly displaced through the 1950s and 1960s and finally fully dispersed by 1970, leaving the city's Black population decentred and scattered through the Metro area. The culprits in Hogan Alley's demise are many, but as is the story with so many Black communities throughout North America, the most obvious villain was a highway.

In the wake of the Great American Highway buildout of the 1950s and 1960s, Vancouver's Modernist growth-machine planners dreamed of a major expressway running through the heart of the city. Vancouver was a small and globally insignificant city in the 1960s, though, and the city's elite were unable to summon enough capital to pull the scheme off; consequently, a determined coalition of residents and activists shut the project down, leaving Vancouver the only major North American city without an inner-city highway. However, one part of the proposed system *did* get built. The Georgia Viaducts were the first phase of the proposed highway and the only part that got completed, sitting orphaned in the middle of the city for four decades. That one little section was constructed directly on top of Hogans Alley, clearing the community out in the name of civic improvement.

Now, however, the Viaducts are going to be removed. There are massive seismic upgrades required, and instead of pouring money into repairs, the City sees a huge opportunity: tear down a relic of past planning disasters and open up a whole section of prime inner-city property for development. The land under the Viaducts – much of it is city-owned – is perfectly located and has massive profit potential, but there is also considerable potential to right a historic wrong.

The Hogan's Alley group of course recognized this before anyone and has been organizing to build something on that land. At first, they were talking about some kind of memorial to remember the historic Black neighbourhood, maybe a plaque or something. Then the group got more ambitious, speaking of a Black community and cultural centre. They are still interested in both of those, but they are now also deep into planning to build a Community Land Trust (CLT) and have received financing and city support to make it happen. It is an amazing, genuinely visionary work, and it may well be able to return Black residents to the area and to some extent reverse a terrible story of displacement.

But the group is also interested in the implications of this work on land that has been subject to multiple expulsions. Hogan's Alley was a Black neighbourhood, for sure, but it was not just that. Every account I have ever read of the place describes it as hoppingly multicultural – with Italians, Chinese, Japanese, and Indigenous people living and working there. Further, Hogan's Alley is just one among many communities in the immediate area to have suffered displacement. The neighbourhood immediately to the north is Chinatown, where Chinese people were historically confined via a long series of official and unofficial racist practices, and it is now being cleared of its Chinese residents and businesses via aggressive gentrification. One more neighbourhood north is Vancouver's Japantown, which during the Second World War saw its community displaced as Japanese residents had their homes and businesses seized and people shipped en masse to internment camps in the interior of the province. Of course, all of that land in question was (and is) Indigenous land, all of it violently seized and occupied over the nineteenth and twentieth centuries.

The claim that Vancouver's Black community has on the land under the Viaducts is undisputed, but so are multiple other claims. The CLT they are proposing is on Indigenous land, amidst significant historical wrongs perpetrated against Japanese and Chinese people. The Hogan's Alley Working Group is more aware of this than anyone, grappling with the questions in front of them and acutely sensitive to how their claims to re-emplacement intersect with other claims. This is partly why I was hanging out with them the other day, to try to think these questions through, and I was really curious to see what you thought of the issue. (I asked for permission to invoke their work here, and the HA working group is willing to have me bring this to our discussion.)

It is almost exactly in our wheelhouse – how to resolve these multiple overlapping claims in the face of multiple, overlapping displacements, and expulsions. There is a significant pressure involved: the timelines for tearing down the Viaducts are accelerating. There are developers and vultures of all kinds circling that land, and there is incredible profit-potential just sitting there. It is a huge and almost-blank slate right in the middle of the city in a highly desirable location. What kinds of thoughts do you have on this?

LKB

I think the instinct to ground this conversation about place and dispossession, reparation, and return in a particular neighbourhood is the right one. The productive working through of these concepts has to happen in real time/real space, and as in the case of Hogan's Alley in Vancouver or Albina in Portland, it has to happen quickly because market forces are not waiting for the theorizing to make moves.

What I am interested in is the potential in Hogan's Alley to address very real material problems of housing and small-business and community spaces, alongside re-telling the story of the neighbourhood in a way that could be degentrifying: if we see gentrification in its metaphorical workings as papering over a neighbourhood's history with a story that is instrumental to capital, can we reconstruct a narrative that resists that? To be more specific, gentrification (and prior to it, urban renewal) requires the framing of a place as a null space, as Glen Coulthard describes. When the viaducts were put in, it was because the neighbourhood was untenable, blighted, perhaps contagiously so. Now, with new development opportunities, some parts of the community history – the cool parts – could be appropriated to jazz up the new scene, maybe by painting a mural or naming a bar after some old thing that is no longer there. Woods (2017) talks about the desecration of sacred spaces as a foundational practice as white wealth encroaches; he also discusses how we might remake sacred spaces through the community's story of everyday life and vitality that also acknowledges enclosure and economic traps, which he calls a 'blues epistemology' of geography.

The idea of cross-cultural storying to build mutual understanding of the distinct and common impacts of white supremacy and capitalist values, to build solidarity among these experiences, seems important in these places where many peoples have lived/been uprooted/held on. Those stories would have to start from indigenous land; they would have to include knowing that the Black Pullman porters were former slaves or a generation away from slavery in the United States or descendants of enslaved people in the Caribbean; they would have to acknowledge 'yellow peril' and its persistence today … As we build a fuller understanding of the layers of dispossession and expropriation that have occurred, as well as resistance moments and failures of solidarity, could we use this not only to revise the narrative of the past but also to develop a new spatial imaginary based on an alternative value set? Could that change the current situation from 'multiple overlapping claims' to a more collective orientation? I am interested to know more about the distinct claims or conflicts arising in this neighbourhood redevelopment opportunity. The architect working on the community centre has said that the Black community has a special claim to the place because their displacement left them uniquely dispossessed – without any other neighbourhood to go to or rebuild as a community. Is that something other folks find legitimate?

MH

Yes, this thread here does make good sense to me. I actually called the architect to discuss this, but I have not reached her yet. (Surely, she has better things to do with her time than engage in semi-obscure theoretical discussions with a rando white lad.) The 'special' claim of 'unique' dispossession seems a little limited, but not untrue. That is, there is more that can be done than instrumentalize that claim in a normative rights-based way. It seems that the visual of 'overlapping' claims actually does a disservice to the idea of 'liberate' – it feels like it might be possible to reach for something more. But her claim is compelling in and of itself.

LKB

In the meantime, since you wrote about Portland in *What a City is For*, there has been some movement on anti-displacement policymaking. The North/Northeast Strategy of the Portland Housing Bureau has kicked off and will be operating with a 'preference policy' that seems to be the first of its kind in the United States (with several cities since adopting similar concepts for their programming). The preference policy, which is more commonly known in the community as 'right to return', gives priority status in subsidized housing programs to low-income families who can show intergenerational residence in Albina. It is tagged to the city's urban renewal actions over time, and the most 'points' are awarded to those whose homes were actually condemned and demolished by the city to build the sports coliseum complex or Legacy Hospital.

I sat on the task force that developed the preference/right to return policy. It has been a really sincere effort by the staff – several of whom are Black Portlanders who grew up in Northeast Portland and have family there or displaced from there – to create a program to prioritize the return of folks who have experienced the greatest impacts of gentrification in Portland. The community members who are on the program's oversight committee are concerned about how the return might go for families who have been living elsewhere, with memories of a tight-knit community in place, and who will return to a completely different social landscape – as well as a lack of low-cost food, changed policing of public space, and neighbours who are living in USD 750,000 houses who are still pretty pissed about that Trader Joe's. The neighbourhoods have gone from being eighty per cent Black to eighty per cent white, the public schools have been realigned, and some social service supports are relocating to East Portland to follow those eligible for their programmes. Thus, the 'returnees' will have a very different landscape to contend with. Claims to 'black space' or the historic heart of a community are recognized in the abstract; if hundreds of Black folks move back in the neighbourhood, there will be real daily contestations over place and representation. How does the homeowner who said that Black people should have nothing to say about the corner of MLK and NE Alberta because 'they don't even live here anymore' react when they do (live there again)? As far as the capital-C City of Portland goes, Northeast is acknowledged as where the Black community is rooted and has history. Whether that is a live, present recognition in real space and with respect to real bodies is a question mark.

Hundreds and hundreds of applications have been made for what is at this point just 140 housing units. The massive response to the programme shows the depth of need for stable and affordable housing among Portland's Black community – within which the average family cannot afford a two-bedroom apartment anywhere in the city. The new buildings are nice enough apartments, but let us be real – when Black folks were talking about restitution and repair for the harms of slavery through urban renewal, we did not mean 'being invited to be on a multi-year wait list to be considered for eligibility for a tiny number of income-restricted units so we can pay our rent to a white-dominated non-profit industrial complex.'

It is probably too much to imagine the City government taking up 'decolonization is not a metaphor' right now, but it is troubling and disappointing for the future of a de-gentrified city that there has been a real failure to move land significantly out of the market. I helped lead an advocacy effort to get land banking as a component of the North/Northeast strategy programming, pushing to get the City to buy land for sale on the market and to expedite the disposition of idle, publicly owned lots, putting both into a new community land trust; we also suggested purchasing existing multi-family housing from owners who hoped to maintain affordable housing in the neighbourhood. Here is where we have run up against the limits of imagination. The routine programme models – giving someone a down payment grant or providing hard equity for a non-profit Community Development Corporation to build apartments – can be adjusted to recognize displaced people, but the models themselves remain static as far as concepts of property. In part, this is because of the many voices from the Black community who are still very much invested in a model of wealth-generating homeownership and are suspicious of a cooperative model that is 'less than' what white folks have had for multiple generations.

It is complicated. As Connolly (2016) points out, twentieth century liberalism has promised that Blacks will move towards citizenship as they approximate whiteness and own property, but that promise has been false. The attenuated forms of land ownership available to Blacks have been economic traps that have also actively diminished self-determination. Nevertheless, there is the ever-present reference to 'forty acres and a mule' – not forty acres that are held in a complicated trust by an entity you do not really understand and that might turn out to have been sharecropping all along. The local CLT organization does not use the language of racial justice, restoration, reparation, or de-gentrification; their pitch sounds to some people like this: 'this "nice white lady" is saying I'm not really going to own my house …'. Recently, we had a visit by Tony Pickett, an African American CLT director who has been working very hard to change this narrative by telling the story of the first CLT in the United States – the New Communities Initiative in Georgia. He told me he is actively trying to dismantle the myth of the origins of CLT as a project of Bernie Sanders in Vermont that is so distant and so white for many folks, to remind/relearn how Black folks actually maintained resilience over time. But this story continues to leave where the land came from as a blank. Thinking practically about what a decolonized/un-propertied community of people and place might look like still does not build a solidarity between Indigenous and Black people.

MH

This is boss work, Lisa, and very heartening even if incomplete and disappointing in some ways (the efforts in Northeast Portland, not your writing). This may sound a little tepid, but these are some amazing steps in better directions, and surely, they are not the end of the Northeast Portland discussion. Each bit of blue sky peeking through the cloud of property heartens others, as long as it is not positioned as just a lifeboat for some, enough to ignore the rest of the ocean. (If you would excuse the laboured analogy.)

I just spent time with this (very) white community land trust in Ontario encouraging them to leverage all their good fortune into something far more than just a safe space for a few nice old White hippies. They are a classic story: fought the good fight for decades, saved this wonderful little community via an innovative CLT, and now they think they have won. And they have, if their goal was a sweet and safe space for a bunch of settled white families. By that standard, their efforts have been a success. But if the horizon is a little different, their success can only be a starting point, not any end. It should be the lever to more and better victories, but it requires them stepping out further, not just hiding in the warm houses and asking how their possession came to be. As Blackstone (1893) put it in the 1780s,

> There is nothing which so generally strikes the imagination, and engages the affections of mankind, as the right of property; or that sole and despotic dominion which one man claims and exercises over the external things of the world, in total exclusion of the right of any other individual in the universe. And yet there are very few, that will give themselves the trouble to consider the original and foundation of this right.

At your suggestion, I just read Justin Leroy's *Black History in Occupied Territory* (2016), and it is an excellent contribution to the conversations around exceptionalism and the entanglements/tension within and between critical Indigenous/Black studies. Among many tremendous lines, this one struck me as particularly insightful: 'The projects of slavery and colonialism have never been concerned with which came first, or which is more elemental – they have in fact thrived on the slippages and ambiguities of their relationship to one another'.

I guess here is one of my (many) questions. I have done a certain amount of reading – not nearly as much as I would like – and I still have a long way to go in my thinking about the intersections and relationships between colonialities and anti-Black ideologies, between slavery and occupation. However, as far as I can tell, critiques of exceptionalism largely conclude by suggesting that scholars and activists should 'bridge the divide', 'learn from one another', 'acknowledge each other's struggles', and so on. These kinds of encouragements/exhortations seem essential but not sufficient. I think Intersectional theory has gone some distance in doing this work regarding de-essentializing class/race/gender et cetera, but I wonder about the particular dissonances between some Black and Indigenous theorizing.

It strikes me as dangerous to conflate or subsume or to gloss over the specificities and particularities and differentiation of struggles. It is also deeply counterproductive to get bound up in questions of which rationality is most terrible, most essential, most foundational, and so on. Leroy's (2016) statement that 'slavery and colonialism have never been concerned with which came first, or which is more elemental' seems really insightful and maybe points in a direction. I wonder then, well, *what are those projects concerned with?*

There are many obvious answers to that question, but some are deceptively slippery to me, and I wonder if that question offers possibilities for points of resistance and organizing. Even more specifically, I wonder if you consider a critique of property useful here – not as the original font of evil, but as one thread to be pulled on. Property seems to be one commonly held rationality – a rationality that can see anything as property (land, water, air, bodies, organs, animals), anything as ownable and sellable – something that both slavery and colonialisms want, or 'are concerned with'.

LKB

Since moving out of the American South, one thing that strikes me is that the historical work that Leroy has done, tracing the intertwined Black and Indigenous history, is less complete here. I think about Woods' (2017) work on the Mississippi Delta region as so valuable to understanding how planning (urban and rural) has maintained the plantation. Moreover, his work is very clear on the appropriate starting point of the region's history being hundreds of years before Europeans appeared. He tells us how the plantation economy required, first, the genocide of Indigenous people and the conversion of land into large property holdings and, then, the labour of chattel slaves to achieve a global capitalist economic profit. He was always careful about 'regionally specific traps' and the need to understand the particulars in a place. Woods compares research and social science to a scalpel – it is the same tool in the hands of the coroner and the surgeon, used for different purposes. How do we talk about history and contemporary analysis in a way that helps the patient to live? This, for me, connects to the problem of Native people being seen as entirely historical and invisible in contemporary urban space, which is especially an issue in Portland, where there's a large urban Indian population (many not officially tribally enrolled because they were born off the reservation or because of the weird politics of which tribes are federally recognized). There is something especially important about both a history with presence and a today with presence. It also means rejecting the rendering of Black communities as 'gone' from gentrifying neighbourhoods, invisibilizing people who continue to live right next door as no longer having a claim to be there. In the organizing work in Portland, the work to understand how settler colonialism sets up a relationship of humanity, land, and property for Indigenous, Black, and settler and how that has worked distinctly here – that has not happened yet. We do get stuck in a less productive space about racialization and minoritization, which minimizes Indigeneity and also skims the surface of the very intense anti-Black history of Oregon as the only state that barred Black residence from its inception.

MH

As I write this, I am literally watching out my window as this outrageously greasy real estate agent works the hell out of some clients, trying to sell the house across the street. It is three floors: two single mothers with kids on the top two floors and a young couple in the basement. Three households. From my window, I can hear him talking about how 'Absolutely, after sale the rents can be raised'. I went downstairs without a shirt on and talked loudly on the porch about this prick's awful predatory bullshit and tried to look sketchy – not a big trouble for me. They glanced over at me a few times, and I tried my best 'stinkface'. I was feeling rather Maoist at the moment, mentally prepping that dude's struggle session.

LKB

Ha ha! My anti-gentrification move is to walk my pit bull around my neighbourhood wearing the biggest hoop earrings I have. I assume the property speculators and class collaborators have heard the old saying 'the bigger the hoops, the bigger the guillotine!' (I am pretty sure that is how it goes.)

Our personal efforts at fashion-based de-gentrification of our immediate surroundings notwithstanding, this dialogue is a first attempt at pulling some threads loose. We have made a start at recognizing that what seems like a theoretical impasse – between Back and Indigenous perspectives in the settler colonial state – must be linked via the common perspective of the making of property. It is a tricky spot – it is hard to find a language away from theft, ownership, appropriation, and reparation – but it is one that feels more feasible than adjudicating claims of who has been the most dehumanized by a system that is not for any of us. I hope we can add more voices to this conversation as we read and think and write our way through.

References

Blackstone, W. (1893) *Commentaries on the Laws of England in four books*. Philadelphia, PA: J.B. Lippincott.

Coates, T-N. (2014). The case for reparations. *The Atlantic*, June.

Connolly, N. D. B. (2016). A Black power method. *Public Books*, June 15. Available at: http://www.publicbooks.org/a-black-power-method/

Coulthard, G. S. (2013). Urbs nullius: Gentrification and decolonization [YouTube video]. Retrieved from: http://ecosocialistsvancouver.org/glen-coulthard-%E2%80%93-urbs-nullius-gentrification-and-decolonization

Desmond, M. (2017). How home ownership became the engine of American inequality. *New York Times Magazine*, May 9.

Douglas, L. (2017). African Americans have lost untold acres of land over the last century. *The Nation*, July 17–24.

Kinney, J. (2017). New Orleans artists take on real estate's loaded terms. *Next City*, May 19.

Leroy, J. (2016). Black history in occupied territory: On the entanglements of slavery and settler colonialism. *Theory and Event*, 19(4), pp. 1–34.

McKittrick, K. and Woods, C. (eds.). (2007). *Black geographies and the politics of place*. Toronto: Between the Lines.

Nichols, R. (2018). Theft is property! The recursive logic of dispossession. *Political Theory*, 46(1), pp. 3–28.

Roberts, A. (2018). Performance as place preservation: The role of storytelling in the formation of Shankleville community's Black counterpublics. *Journal of Community Archaeology and Heritage*, 5(3), pp. 146–165.

Simpson, L. B. (2013). Restoring nationhood: Addressing land dispossession in the Canadian reconciliation discourse. Lecture – SFU's Vancity Office of Community Engagement, November 13th.

Tuck, E. and Wayne Yang, K. (2012). Decolonizatio is not a metaphor. *Decolonization Indigeneity, Education and Society*, 1(1), pp. 1–40.

Tutashinda, C. (2017). Reparations for Black land and liberation manifesto. *Black land and liberation*, May 24. Available at: http://blacklandandliberation.org/uncategorized/manifesto/

Woods, C. (2017). *Development arrested: The blues and plantation power in the Mississippi Delta*. 2nd ed. NYC: Verso Books.

Conclusion

Housing displacement: conceptual and methodological issues

Maria Persdotter, Guy Baeten, Carina Listerborn, and Emil Pull

The displacement of low-income, racialized, and otherwise marginalized households is an enduring feature of capitalist urban development in much of the world. Researching displacement – in terms of its incidence, causes, mechanisms, and consequences – is thus imperative to expose and challenge urban injustice. It has the potential to flip the script of many urban redevelopment programs, premised as they often are on promises of neighbourhood improvement, and provide those resisting such programs with a powerful moral and political argument: displacement from home and neighbourhood destroys people's attachments to place and to each other – so much so that its negative social and health impacts can sometimes be quantitively measured (Fullilove, 2016; see also the discussion in Hubbard and Lees, Chapter 1).

Over the past several years, there has been a renewed interest in displacement and forced evictions within the fields of housing and urban studies. Nevertheless, the conceptual and methodological literature on displacement remains marked by a certain Anglo-American bias (Maloutas, 2011, 2018). Therefore, when we first decided to put together this edited collection, we wanted to include a diversity of case studies and to think across geographic and conceptual divides. In particular, we wanted to untangle the concept of displacement from the dominant frameworks of Anglo-American gentrification theory. In closing the collection, we wish to conclude with four final reflections on researching housing-related displacement. The first addresses the place of gentrification theory in research on contemporary housing displacement and injustice, the second concerns the temporalities and historical roots of displacements, the third focuses on methodological challenges – and opportunities – in researching housing displacement, and the fourth relates to the recent 'phenomenological turn' in displacement research.

Untangling 'displacement' from the frameworks of gentrification theory

Arguably, gentrification categorically *is* displacement (cf. discussion in Lees et al., 2016: 36). However, as several of the chapters in this collection demonstrate, the opposite is not always the case – housing displacement is not always gentrification (see Pull, Chapter 3; Baeten and Listerborn, Chapter 7). Gentrification is

> a process involving a change in the population of land-users such that the new users are of a higher socio-economic status than the previous users, together with an associated change in the built environment through a reinvestment in fixed capital.
>
> (Clark, 2005: 258)

To be sure, gentrification is still a highly relevant concept to contextualize and explain the displacements of low-income residents in many settings. The collection features a number of studies of gentrification-induced displacements, including Melissa García-Lamarca, James Connolly, and Isabelle Anguelovski's quantitative analysis of 'green gentrification' in Barcelona, Spain (Chapter 10) and Anne-Sophie Iotti's analysis of state-led gentrification in Brisbane, Australia (Chapter 9).

While we are sympathetic to the argument that gentrification is a 'global' or 'planetary' urban strategy (Smith, 2002; Lees et al., 2016), we also maintain that gentrification, as a concept, is insufficient to capture the diverse causes and mechanisms of housing displacements today. Based on a case study from Uppsala, Sweden, Emil Pull (Chapter 3) suggests that Swedish cities are currently experiencing a paradigmatic shift from market-driven gentrification to a new form of accumulation by dispossession. Across the country, selective and strategic renovations of the 1960s and 1970s rental housing stock are driving up rents and creating significant displacement pressures. In Uppsala, rents have increased by as much as sixty-five per cent in some neighbourhoods. Contrary to what gentrification theory would have us expect, residents are not being physically pushed out and replaced by wealthier residents. In fact, average income levels in the affected neighbourhoods have remained more or less stagnant even as rents have surged. According to Pull, many low-income residents have been forced to move out of their neighbourhoods due to renovation-related rent increases, but they have not been replaced by residents of a higher socio-economic status. Instead, other low-income families have moved into the newly renovated apartments. Some established residents have also opted (perhaps due to a lack of alternative options) to stay put and absorb considerable rent increases. Thus, Pull concludes that there is a form of gentrification without gentrifiers underway in Uppsala. The rental housing companies are successfully closing a rent gap, but not through the influx of more affluent residents; they are simply extracting more profits from the existing (low-income) tenants. This can perhaps be characterized as a form of indirect displacement.

Another important stake in the conceptual and theoretical debate about gentrification has to do with the role and coercive powers of the state. Asher Ghertner (2015: 553) argues that gentrification theory 'fails in much of the world' because it assumes that private property is 'already the exclusive, or predominant, form of tenure in the places it seeks to describe'. By failing to engage with the heterogeneity of property regimes across the globe, gentrification theory commonly renders 'unthinkable and invisible' the often state-led processes of land transformation that 'underpin the most violent forms of displacement' taking place in cities in the Global Southeast (Ghertner, 2015: 552). Jonathan McCombs' study of post-socialist gentrification in Hungary's Eight District, Budapest, lends some support to this argument. McCombs explains that the post-socialist transition and privatization has produced a super-homeownership property regime, where a majority of the Hungarian population own their apartments. The structure of this property regime, McCombs notes, 'limits the impact of the market on gentrification processes, because there is not a concentration of property in the hands of a developer class, that can induce mass displacement'. The gentrification of the Eight District and other similar neighbourhoods is instead a state-driven process that revolves around the biopolitical management of people and place. The renewal of the Eight District (and the displacements associated with it) has depended, and continues to depend, on the prior problematization of the area as needing intervention due to the allegedly poor conditions of housing and the relatively high concentration of Romani households found there. McCombs suggests that the targeted evictions of Roma is envisioned as a strategy to physically and socially rehabilitate the neighbourhood and concludes that displacement and gentrification in the Eighth District 'should be conceived as fundamentally biopolitical' – a process which depends in large part on the coercive powers of the state (Chapter 5).

McCombs' analysis resonates with that of Louise Fabian and Anders Lund Hansen (Chapter 8). Focusing on a pre-eminently biopolitical technology of government – the Danish government's so-called ghetto plan – Fabian and Lund Hansen argue that the state-led territorial stigmatization of immigrant-dense neighbourhoods plays a crucial role in the ongoing commodification of the Danish housing commons. According to Fabian and Lund Hansen, the ghetto plan marks the latest in a series of attempts to weaken the non-profit housing sector and open it up to the White and wealthy (upper) middle classes. The chapter emphasizes the coercive nature of state-intervention in this process; whereas ghettos have historically been used to contain racialized minority groups (especially Jews), the Danish government is using the stigmatizing language of 'the ghetto' to enforce an aggressive policy of compulsory dispersal.

If gentrification is commonly understood as a market-driven process in which the state plays a secondary role, Guy Baeten and Carina Listerborn (Chapter 7) call attention to the (re)-emergence of direct forms of exclusionary banishment. They point to a shift from urban 'boosterism' strategies that aim to attract high-income earners, capital, and wealth to the city to policies that aim to directly exclude the poor. To illustrate their argument, they draw on a case study from Landskrona, a small municipality in the south of Sweden that is experimenting with highly restrictive rental policies in an explicit attempt to deter low-income households from settling in the municipality. The ultimate aim of the policy is to reduce municipal public spending on social assistance and re-make Landskrona as a middle-class, suburban municipality. Crucially, banishment strategies (such as the Landskrona rental policy) rely not on the mechanisms of the market but on the direct coercive and prohibitive powers of the state. Through a diverse repertoire of tactics, the municipality aims to shift the social composition of their population and curb the influx of benefit-dependent citizens.

The temporalities and historical roots of displacement

The chapters in the collection show that housing displacement, as David Delaney puts it, is 'less a singular event than an ongoing *state of being*': a condition defined by the often-violent experience of repeatedly being 'pushed around, expelled, removed ... given the bum's rush, kicked out' (Delaney, 2004: 848, our emphasis). Although displacement can be a singular act (as in the case of mass demolitions), more often it occurs 'through a series of smaller aggressions' that displace residents and businesses over a longer time-span (Elliott-Cooper et al., 2019: 10). Indeed, several of the chapters in the collection demonstrate that displacement does not begin and end at the moment of, say, an eviction. Often displacement pressures (Marcuse, 1986) build incrementally over an extended period of time in a way that causes established residents to feel profoundly 'out-of-place' long before being – and even if they are never – physically dislocated (see Hubbard and Lees, Chapter 1; Iotti, Chapter 9; Pull, Chapter 3). Moreover, displacement does not end when one decides to relocate or is resettled elsewhere; the negative individual psychological impacts and social consequences of displacement are significant and long-lasting. Ipsita Chatterjee's study of the mass displacement caused by the redevelopment of the Sabarmati riverbanks in Ahmedabad, India, aptly illustrates this point (Chapter 2; see also Chatterjee, 2014). Chatterjee argues that the threat of displacement is always present in the lives of the urban poor, so much so that their displacement comes to appear as inevitable. The communities living along the Sabarmati river had first ended up moving there as a result of previous cycles of dispossession and displacement. Once displaced from the location, they were resettled to a formerly industrial site on the outskirts of Ahmedabad, which contributed to their further marginalization; it cut them off from livelihood opportunities and served to exacerbate inter-communal conflict and violence. Chatterjee (in Chapter 2) thus concludes that 'displacement is not the beginning and end of exploitation'.

Indeed, any act of displacement builds upon and reinforces pre-existent power relations predicated on class, race, and gender. Hence, there is a need to situate contemporary displacements within geographically and historically specific gendered and racialized formations of class exploitation and (post)colonial dispossession. Just as we cannot understand the displacement of the inhabitants of the Sabarmati area without grappling with the complex geographies of class and inter-communal Hindu-Muslim relations in Gujarat, India, we cannot make sense of the unfolding transformation of Budapest and other cities in Eastern Europe without considering how the post-socialist transition and the emergence of a 'super homeownership' housing regime intersects with the region's particular history of racialized dispossession to entrench the segregation and subjugation of the Romani minority (McCombs, Chapter 5). Similarly, we cannot properly understand the gentrification-driven displacement and marginalization of Black and Indigenous communities from North American cities today without considering that the very institution of property ownership is indelibly shaped by the entwined history of settler colonialism and slavery – what Matt Hern and Lisa Bates (Chapter 11) refer to as the originary twin thefts of people from land and land from people (see also McElroy, Chapter 6).

The difficulties of evidencing displacement

While the contributors to this collection employ a variety of different methods – ranging from quantitative spatial analysis, via community-based research and mapping of evictions, to qualitative interview-based and ethnographic methods – most seem to agree on the difficulty of charting and evidencing displacement. Indeed, attempting to quantify displacement presents us, as researchers, with significant methodological challenges. Given what we know about the complex dynamics of displacement pressures (Marcuse, 1986), what is the appropriate timeframe for measuring displacement? How do we devise a methodology that is able to simultaneously capture direct, short-term displacements (e.g., mass evictions) and longer-term, indirect ones? How do we go about distinguishing between forced and voluntary mobility? And how do we capture the destructions of people's attachments to place that can happen when a neighbourhood changes without there being much, or any, physical out-migration?

Phil Hubbard and Loretta Lees (Chapter 1) describe just how difficult and time-consuming it can be to find relevant data on displacement. They set out to measure and map the displacements of residents from London's gentrifying council housing estates only to discover that the range of data available for actually tracking the displaced was 'frustratingly limited'. The unavailability of relevant data is perhaps itself a sign of neglect on the part of the authorities. Hubbard and Lees (Chapter 1) point out that

> in many ways, the lack of data on the numbers displaced has allowed governments, policy-makers and planners to pursue strategies of gentrification unchallenged by statistical evidence of what is often mooted as its most negative impact: the displacement of long-term residents.

These problems are, of course, even more pronounced in the case of informally or precariously housed residents. Even when there is public data to measure out-migration of residents, those residents who have no official lease or who are not formally registered on a given address are typically completely invisible in official statistics. Thus, there is a sense in which not being entirely 'legible' (Scott, 1998) in the eyes of the state makes an individual or community even more radically displaceable (cf. discussion in Persdotter, 2019: 165). Undocumented migrants are, perhaps, the paradigmatic example (Keenan, 2017), but there are also other groups of precariously housed and partially 'undocumented' residents that are largely invisible in official statistics related to displacement. For example, a study of evictions in Sweden concluded that the displacements (broadly defined) of families on social assistance are significant in number and escalating. However, such displacements are almost entirely invisible in the public eviction records. This is largely because families on social assistance often sublet their apartments from the municipality. Thus, their evictions typically do not register as 'evictions' (in the narrow, technical-legal meaning of the word) but as a quotidian administrative decision (Rädda barnen, 2017; Samzelius, 2018). This is a problem insofar as it leads to an underreporting of *de facto* evictions. Furthermore, it should tell us something about how the dynamics of displacement intersect with the precariazation of life and housing (Vasudevan, 2017; Listerborn, 2018).

In light of the inherent challenges involved in measuring displacement, several of the chapters in this collection argue for the need to supplement quantitative data with qualitative approaches. For instance, Hubbard and Lees (Chapter 1) conclude that even if it would have been possible to track the flows of displaced people from the council estates, a quantitative approach is still insufficient to account for indirect forms of displacement:

> The displacement associated with the renewal of London's council estates is rarely the product of a one-off event. More often it is a form of slow, percolating violence. Displacement is not just about direct replacement of poorer groups by wealthier ones; it also involves forms of social, economic, and cultural transition that alienate established populations. This can entail forms of slow violence that render particular neighbourhoods less hospitable and accommodating to established residents over time, as well as direct and forceful acts of expropriation which the vulnerable and precarious seem least able to cope with. In this regard, in addition to quantitatively tracking displacement itself, it appears vital that we evolve better qualitative measures of displacement pressures in order to predict, and ideally help prevent, future displacements.

Similarly, Melissa García-Lamarca, James Connolly, and Isabelle Anguelovski (Chapter 10) emphasize the need to complement quantitative longitudinal and spatial analyses of green gentrification-induced displacements with qualitative research in order to identify the cultural, social, and procedural dynamics of displacement.

Displacement and housing researchers frequently engage in scholar activism (Driscoll Derickson and Routledge, 2015). This is reflected in housing journals such as *Critical Housing Analysis* and *Radical Housing Journal* and in the proliferation of activist-oriented handbooks such as *Staying Put: An Anti-Gentrification Handbook for Council Estates in London* (London Tenants Federation et al., 2014). Scholar activism is also at the heart of Critical Urban Sustainability Hub (CRUSH), the research platform of which the editors of this book are part. The Anti-Eviction Mapping Project (AEMP) – a data visualization, critical cartography, and storytelling collective based in the San Francisco Bay Area, USA – shows that community-based research can be a critically important tool in the struggle against displacement and gentrification. In Chapter 6, Erin McElroy, one of the founders of the AEMP, reflects on the organization's efforts to identify the actors behind Limited Liability Corporations (LLCs). According to McElroy, LLCs have become major players in the financialized Bay Area housing market. These corporations are used by landlords to amass financial gain while providing a form of strategic anonymity to speculative evictors. By identifying the actors behind LLCs, the AEMP and their partner organizations have been able to wage effective campaigns against some of the companies responsible for the unfolding eviction crisis in the area. Elroy's chapter speaks to the importance of taking resistance seriously.

Indeed, if displacement is a condition marked by a profound sense of powerlessness, this is not to say that displacees do not have agency. As several of the chapters in the collection show, the threat of displacement can be a powerful galvanizing force to action (McElroy, Chapter 6; Valli, Chapter 4). In many cases, it brings together neighbours to generate communities of resistance, and sometimes those communities are successful in their efforts to stay put. According to Brickell et al. (2017: 18), '[a]ctivist practices [challenge] the inevitability of displacement and expulsion while producing new resistant geographies that remap the traditional boundaries between the intimate and the political, and the local and the global'. Louise Fabian and Anders Lund Hansen's discussion (Chapter 8) on the growing resistance to the Danish government's so-called ghetto plan lends support to this argument. They show how neighbourhood organizations are challenging and subverting the racializing and stigmatising discourse of the 'ghetto policies', in part by re-appropriating the very term 'ghetto'. According to Fabian and Lund Hansen, new alliances are being formed in opposition to the 'ghetto plan' that are reinvigorating the public debate on the contemporary housing question. A new generation of housing activists are exploring new repertoires of action to mobilize resistance to marginalization and stigmatization and to fight for the right to freedom from displacement.

Emotional and experiential dynamics of displacement

Displacement is a traumatic experience, one that frequently elicits feelings of anger, frustration, grief, and uncertainty (see e.g., Pull, Chapter 3; Valli, Chapter 4). Several of the chapters in this collection advocate for a methodological approach that starts from the experiences of displacees (see, in particular, Chatterjee, Chapter 2; Valli, Chapter 4). This is important as the experiential dimension of displacement easily gets lost in quantitative measures of displacement as physical dis/re-location of individuals and households (i.e., out-migration). By treating displacement as a purely spatial event, we are surely missing something of its intensely felt traumatic aspects. Arguably, we are missing the very reason why displacement matters in the first place (cf. Davidson, 2009: 223; see also Fullilove, 2016).

Responding to the inadequacy of the dominant conceptualization of displacement as physical dis/re-location (or 'out-migration'), several scholars have turned to phenomenology to re-theorize displacement as the destruction of people's lived attachments to place and home – what some call 'un-homing' (Atkinson, 2015; Elliott-Cooper et al., 2019). For instance, Mark Davidson (2009; 226, emphasis in original) argues for the need to 'reassert the place in dis*place*ment'. According to Davidson, the ongoing scholarly debate on gentrification-induced displacement is continually hampered by the predominance of the 'abstraction of displacement-as-out-migration' – an abstraction which, in turn, is based on a Cartesian understanding of space as a mere container for social processes. Davidson (2009: 226) draws on Martin Heidegger's place-based ontology of being and Henri Lefevbre's work on place as a form of 'lived space' to argue that any analysis of displacement must take into consideration the subjective experience of place and the ways in which it anchors identity and existence. In other words, to understand what is lost when a neighbourhood undergoes gentrification requires us to engage with the many dimensions of 'lived space' and with the micro-political meanings of the experiencing of the un-making of community and home spaces.

Taking the phenomenological – experiential and embodied emotional – dimension of displacement seriously compels us to rethink our established methods for charting and measuring displacement. It also opens up novel and potentially productive approaches. Decoupling 'abstract space' and 'lived space' allows us to grasp conceptually how 'people can be displaced – unable to (re)construct place – without spatial dislocation, just as much as they can with spatial dislocation' (Davidson, 2009: 228). Furthermore, 'the diagnosis and conceptualisation of gentrification-induced displacement as *a form of severance* allows us to look for different signs of displacement, with affected neighbourhoods and populations displaying the marks of wounding or trauma' (Elliott-Cooper et al., 2019: 7, our emphasis; see also Hubbard and Lees, Chapter 1).

Crucially, an attention to the affective, emotional, and experiential dimensions of displacement offers important clues to the 'how' of displacement. Expanding on Marcuse's (1986) early theorization of 'displacement pressure', Chiara Valli (Chapter 4) posits displacement as an affectively charged process that revolves around the symbolic violence of devaluation, misrecognition, and Othering. According to Valli, the symbolic violence directed (intentionally or unintentionally) at established predominantly Black and Brown Bushwick (NY, USA) residents by wealthier and Whiter newcomers to the neighbourhood is not just an effect but a driving-force of gentrification in the area. It erodes long-term residents' sense of place and belonging and elicits feelings of anger, frustration, and grief. Valli proposes the term 'emotional displacement' to capture the process whereby an individual or community come to experience a profound 'loss of place' even without being physically dis/re-located. In Valli's words (Chapter 4), 'the intangible components of displacement – "cultural", "everyday", "discursive", or "emotional", and more – are significant and critical expressions of the power structures at play in urban renewal processes'.

There is much to be said about the phenomenological turn and its emphasis on place as an anchoring-point for identity and community. It provides an important corrective measure to the overly simplistic conceptualizations of space that underlie most quantitative measures of displacement. At a moment in time when mobility and unattachment are widely celebrated as emblems of freedom, it also serves as a reminder of the value of being able to choose to stay put. Another important merit of phenomenological approaches that focus on people's attachment to place is that such approaches (at least theoretically) have the potential to recognize – and valorize – relationships to place that are not officially sanctioned as home ownership or tenure. This is particularly important for unhoused, property-less, and perpetually displaced subjects, like street-homeless people (cf. discussion in Herbert and Beckett, 2010).

That said, approaches that focus on people's lives, everyday experiences of attachment to place come with certain limitations and (potentially) troubling political implications. For one thing, such approaches are not by definition more inclusive than other ones. If we attempt to operationalize a definition of displacement as the severance of people's attachments to place, we are immediately faced with the question, what counts as a meaningful attachment to place? If our conceptualizations of – and political arguments against – displacement are premised on the subject's prior attachment to a given place, then what about those who, for various reasons, are unable to claim a significant attachment to any one place? If you first have to be 'rooted' to be recognized as a (legitimate) sufferer of 'uprooting', then this is a conceptualization of displacement that leaves out those who were never able to lay down roots in the first place.

As Davidson (2009) himself notes, there is a close – if not determinate – relation between the place-based philosophies of (amongst others) Heidegger and reactionary politics (for deeper discussion of this, see Harvey, 1989: 209; Malpas, 2006; Massey, 2012). Anthropologist Liisa Malkki (2004) refers to the naturalization of the relationship of people to place as a core aspect of a 'sedentarist metaphysics' – a structure of thought that territorializes identities in property, region, and above all, nation (see also Cresswell, 1997, 2006). According to Malkki (2004), the adverse consequences of a sedentarist metaphysics are severe for (involuntarily) mobile subjects like refugees. This is because 'sedentarist assumptions about attachment to place leads us to define displacement not as a fact about socio-political context, but rather as an inner, pathological condition of the displaced' (Malkki, 2004: 33). Hence, just as the solution to dispossession cannot simply be (re-) possession (Butler and Athanasiou, 2013; see discussion in Hern and Bates, Chapter 11), the solution to displacement also cannot be a reactionary defence of place-as-is. This does not offer much hope for a wider politics. As Massey (1992, 2005, 2012) would say, resistance to displacement (scholarly and otherwise) needs to embrace the irreducible 'throwntogetherness' and persistent openness of place.

In summary then, *Housing Displacement: Conceptual and Methodological Issues* offers critical insights into the multi-dimensional processes and experiences of displacement. The collection highlights the need to think beyond the dominant conceptualization of displacement as physical dis/re-location of individuals and households. Displacement fundamentally involves the loss of one's attachments to place. This need not include physical displacement. In fact, a single-eyed focus on physical displacement impedes our understanding of the important emotional and experiential dynamics of displacement. Although displacement is a global phenomenon, displacement processes cannot be properly understood outside their spatial and temporal contexts, which structure the dynamics of displacement. By situating contemporary displacements within geographically and historically specific gendered and racialized formations of class exploitation and (post) colonial dispossession, we can begin to grasp the power-relations behind housing displacements and formulate counterstrategies to empower those who fight it.

References

Atkinson, R. (2015). Losing one's place: Narratives of neighbourhood change, market injustice and symbolic displacement. *Housing, Theory and Society*, 32(4), pp. 373–388. https://doi.org/10.1080/14036096.2015.1053980

Brickell, K., Fernández Arrigoitia, M., and Vasudevan, A. (eds.). (2017). *Geographies of forced eviction: Dispossession, violence, resistance.* London: Palgrave Macmillan.

Butler, J. and Athanasiou, A. (2013). *Dispossession: The performative in the political.* Cambridge: Polity.

Chatterjee, I. (2014). *Displacement, revolution, and the new urban condition: Theories and case studies.* SAGE Publications India.

Clark, E. (2005). The order and simplicity of gentrification – A political challenge. In: Atkinson, R. and Bridge, G. (eds.) *Gentrification in a global context: The new urban colonialism*. London: Routledge, pp. 256–264.

Cresswell, T. (1997). Weeds, plagues, and bodily secretions: A geographical interpretation of metaphors of displacement. *Annals of the Association of American Geographers*, 87(2), pp. 330–345. https://doi.org/10.1111/0004-5608.872056

Cresswell, T. (2006). *On the move: Mobility in the modern western world*. New York: Routledge.

Davidson, M. (2009). Displacement, space and dwelling: Placing gentrification debate. *Ethics, Place and Environment*, 12(2), pp. 219–234. https://doi.org/10.1080/13668790902863465

Delaney, D. (2004). Tracing displacements: Or evictions in the nomosphere. *Environment and Planning D: Society and Space*, 22(6), pp. 847–860. https://doi.org/10.1068/d405

Driscoll Derickson, K. and Routledge, P. (2015). Resourcing scholar-activism: Collaboration, transformation, and the production of knowledge. *The Professional Geographer*, 67(1), pp. 1–7.

Elliott-Cooper, A., Hubbard, P., and Lees, L. (2019). Moving beyond Marcuse: Gentrification, displacement and the violence of un-homing. *Progress in Human Geography*. https://doi.org/10.1177/0309132519830511

Fullilove, M. T. (2016). *Root Shock: How tearing up city neighbourhoods hurts America, and what we can do about it*. New York: New Village Press.

Ghertner, D. A. (2015). Why gentrification theory fails in 'much of the world'. *City*, 19(4), pp. 552–563. https://doi.org/10.1080/13604813.2015.1051745

Harvey, D. (1989). *The condition of postmodernity* (Vol. 14). Oxford: Blackwell.

Herbert, S. and Beckett, K. (2010). 'This is home for us': Questioning banishment from the ground up. *Social & Cultural Geography*, 11(3), pp. 231–245. https://doi.org/10.1080/14649361003637661

Keenan, S. (2017). A border in every street. Available at: https://thedisorderofthings.com/2017/06/29/a-border-in-every-street

Lees, L., Shin, H. B., and López-Morales, E. (2016). *Planetary gentrification*. Cambridge: Polity.

Listerborn, C. (2018). *Bostadsojämlikhet: röster om bostadsnöden*. Stockholm: Premiss.

Malkki, L. (2004). National Geographic: The rooting of peoples and the territorialization of national tdentity among scholars and refugees. *Cultural Anthropology*, 7(1), pp. 24–44. https://doi.org/10.1525/can.1992.7.1.02a00030

Maloutas, T. (2011). Contextual diversity in gentrification research. *Critical Sociology*, 38(1), pp. 33–48.

Maloutas, T. (2018). Travelling concepts and universal particularisms: A reappraisal of gentrification's global reach. *European Urban and Regional Studies*, 25(3), pp. 250–265.

Malpas, J. (2006). *Heidegger's topology: Being, place, world*. Cambridge, MA: The MIT Press.

Marcuse, P. (1986). Abandonment, gentrification, and displacement: The linkages in New York City. In: Lees, L., Slater, T., and Wyly, E. (eds.) *The gentrification reader*. London: Routledge, 2010, pp. 333–347.

Massey, D. (1992). A place called home. *Housing, Theory and Society*, pp. 128–138. https://doi.org/10.1080/14036090410021360

Massey, D. (2005). Throwntogetherness: The politics of the event of place. In: *For space*. London: Sage, pp. 149–162.
Massey, D. (2012). Power-geometry and a progressive sense of place. In: Jon Bird (ed.) *Mapping the futures*. London: Routledge, pp. 75–85.
Persdotter, M. (2019). *Free to move along: On the urbanisation of cross-border mobility controls – A case of Roma 'EU migrants' in Malmö, Sweden*. PhD dissertation, Malmö University and Roskilde University, Malmö/Roskilde.
Rädda barnen. (2017). En plats att kalla hemma: Barnfattigdom i bostadskrisens skugga. Available at: https://www.raddabarnen.se/globalassets/dokument/rad--kunskap/arbetar-for-barn/barnfattigdomsrapporter/barnfattigdomsrapport_enplatsattkallahemma_2017_radda_barnen.pdf
Samzelius, T. (2018). Ensamstående mammor och rätten till bostad. In: Dahlstedt, M. and Lalander, P. (eds.) *Manifest – för ett socialt arbete i tiden*. Lund: Studentlitteratur.
Scott, J. C. (1998). *Seeing like a State: How certain schemes to improve the human condition have failed*. New Haven, CT: Yale University Press.
Smith, N. (2002). New globalism new urbanism: Gentrification as global urban strategy. *Antipode*, 34(3), pp. 427–450.
Vasudevan, A. (2017). Zwangsräumungen in Berlin: Towards an historical geography of dispossession. In: *Geographies of Forced Eviction*. London: Palgrave Macmillan, pp. 191–213.

Index

Note: **Bold** page numbers refer to tables; *italic* page numbers refer to figures and page numbers followed by "n" denote endnotes.

For Product Safety Concerns and Information please contact our EU representative GPSR@taylorandfrancis.com
Taylor & Francis Verlag GmbH, Kaufingerstraße 24, 80331 München, Germany

www.ingramcontent.com/pod-product-compliance
Ingram Content Group UK Ltd.
Pitfield, Milton Keynes, MK11 3LW, UK
UKHW022130190926
475702UK00013B/1725
9780367564537

For Product Safety Concerns and Information please contact our EU representative GPSR@taylorandfrancis.com
Taylor & Francis Verlag GmbH, Kaufingerstraße 24, 80331 München, Germany

www.ingramcontent.com/pod-product-compliance
Ingram Content Group UK Ltd.
Pitfield, Milton Keynes, MK11 3LW, UK
UKHW022123190926
475702UK00013B/1609
9781138278905